AIChE Equipment Testing Procedure

SPRAY DRYERS

A Guide to Performance Evaluation

Second Edition

Prepared by the

Equipment Testing Procedure Committee

American Institute of Chemical Engineers
3 Park Avenue, New York, NY 10016-5991

3 Park Avenue, New York, NY 10016-5991
www.aiche.org
ISBN 0-8169-0925-3
Pub. E-32

Cover and Layout by Beth Shery Sisk

Approved for publication by AIChE's Chemical Engineering Technology Operating Council in Spring 2003. Second edition approved for publication by AIChE's Board of Directors in 2003.

AMERICAN INSTITUTE OF CHEMICAL ENGINEERS EQUIPMENT TESTING PROCEDURE COMMITTEE

Chair: **Dr. P.C. Gopalratnam, P.E.**
DuPont Apparel & Textile Sciences

Vice Chair: **Richard P. O'Connor**
R.P. O'Connor and Associates

SPRAY DRYERS
Procedure Revision Subcommittee

Chairman: **Dr. Richard H. Snow, P.E.**, Chicago Chem Consultants Corp.
General Committee Liaison: **Joseph F. Hasbrouck, P.E.**,
Hasbrouck Engineering, P.C.

Members:

Robert E. Barnes	Engelhard Corporation
Miguel Castro	Hughes Petroleum
Edward M. Cook	Energy Saving Consultants
Mark Roisum	Nyro, Inc.
Paul Schmidtchen	Martin Marietta Magnesia Specialties, Inc.

Members Participating in First Edition

J.J. Wettstein
G.W. Baldwin
R.W. Byliss

S.S. Grossel
David Lee
P.Y. McCormick

First Edition approved by AIChE Council in 1988 for publication.

AMERICAN INSTITUTE OF CHEMICAL ENGINEERS EQUIPMENT TESTING PROCEDURE COMMITTEE

GENERAL COMMITTEE
(2003)

Dr. P.C. Gopalratnam, P.E., *Chairman*
DuPont Apparel & Textile Sciences

Richard P. O'Connor, *Vice Chairman*
Thomas H. Yohe, *Past Chairman*
Prabir K. Basu, *Council Liaison*
Stephen R. Smith, *Staff Liaison*
Beth Shery Sisk, *Staff Liaison*

GENERAL COMMITTEE MEMBERS

Prashant Agrawal
S. Dennis Fegan
Joseph F. Hasbrouck, P.E.
Dr. John G. Kunesh
Robert E. McHarg
Anthony L. Pezone
Rebecca M.W. Starkweather

Table of Contents

100 Purpose and Scope 1
101 Purpose 1
101.1 Purpose 1
101.2 Summary 1
101.3 Scope of Spray Drying 2
101.4 Reasons for Testing Spray Dryers 2
102 Design vs. Operational Variables 5
103 Liability 5
200 Definitions and Descriptions of Terms 5
201 Dryer Design 5
201.1 Dryer Chamber 5
201.2 Airflow Patterns 5
201.3 Product Flow 6
201.4 Atomizer 6
201.5 Heating Methods 6
201.6 Product Recovery 6
201.7 Airflow Motive Force 8
202 Description of Terms 8
202.1 Drying 8
300 Test Planning 11
301 Preliminary Objectives 11
301.1 Test Objectives 11
301.2 Organizational Resources 11
301.3 Schedule 11
301.4 Dryer Controls and Instrumentation 12
301.5 Peripheral Equipment 12
301.6 Pretest Calculations 12
301.7 Test Plan 12
301.8 Environmental 15
301.9 Cleaning and Inspection 15
301.10 "Dry Run" 15
302 Types of Test 15
302.1 Dryer System Capacity 15
302.2 Heat and Material Balance 16
302.3 Product Properties 19
302.4 Acceptance Test 19
303 Variables Affecting Product Properties 19
303.1 Residual Moisture Content 19
303.2 Atomization 20
303.3 Heat Sensitivity 23
303.4 Physical Properties 23
303.5 Prediction of Capacity and Rate Effects 26
303.6 Summary 26
304 Data Requirements - Product Properties 27
304.1 Data Requirements 27
304.2 Atomizer Data 28
304.3 Complete Heat and Material Balances 28
304.4 Peripheral Equipment Limitations ... 29
304.5 Product Quality Measurements 29
304.6 Particle Properties 29
305 Test Preparation 29
305.1 Objective of Tests 29
305.2 Operating Variables 30
305.3 Test Data Sheet 30
305.4 Measurement Methods 30
305.5 Test Plan 30
305.6 Planning Check List 30
400 Methods of Measurement and Sampling 30
401 Gas Temperature and Humidity 30
401.1 Selection of Temperature Sensors ... 30
401.2 Installation of Temperature Sensors 31
401.3 Duct Temperature and Velocity Profiles 31
401.4 Accuracy of Dry Bulb Temperatures 31
401.5 Atmospheric Humidity 31
401.6 Dryer Exit Humidity 31
401.7 Accuracy of Gas Wet Bulb Temperature 31
402 Gas Flow 32
402.1 Installed Flow Meters 32
402.2 Inlet Gas 32
402.3 Exit Gas Measurement 33
403 Material Temperature and Moisture Content 34
403.1 Product Temperature Measurement. 34
403.2 Product Moisture Content 34
404 Dust Flow Measurements 34
405 Radiation and Convection Heat Losses 35
405.1 Estimated Heat Loss 35
405.2 Measuring Heat Loss 35
405.3 Outdoor Equipment 35
406 Miscellaneous Measurements 36
406.1 Static Pressure 36
406.2 Location of Pressure Sensors 36
406.3 Electric Power Measurements 36
500 Test Procedure 36
501 Plant-Scale Test 36
502 Exploratory Experiments 37
502.1 Production Capacity 37
502.2 Product Quality 38
503 Preliminary Trial 38
503.1 Water Run 38
503.2 Dryer Operability 38
504 Definitive Test 39
504.1 Test Start-Up 39
504.2 Running a Definitive Test in a Plant-Scale Dryer 39

505 Humidity and Moisture Measurements 40
600 Computation of Results 40
601 Nomenclature 40
601.1 Variables 40
601.2 Properties 40
601.3 Subscripts 40
602 Material Balances 40
602.1 Dry Solids 40
602.2 Moisture Balance 42
602.3 Psychrometric Chart Method 42
602.4 Humid Air Volume 44
602.5 Total Air Volume 44
602.6 Air Material Balance 44
603 Heat Balances 44
603.1 Enthalpy or Heat Balances on the Dryer 44
603.2 Heater Balances 46
604 Thermal Efficiency 48
604.1 Calculation 48
604.2 Optimization 49
700 Interpretation of Results 50
701 Exploratory or Troubleshooting Tests 50
702 Material, Moisture, and Energy Balances 50
703 Dryer Heat Losses and Thermal Efficiency 50
703.1 Heat Losses 50
703.2 Thermal Efficiency 51
800 Appendix 52
801 Nomenclature 52
802 Example: Evaluating Drying Capacity of Air Stream 53
803 Example: Dryer Test with Psychrometric Chart Heat Balance .. 55
803.1 Known Inputs 55
803.2 Evaluate Product and Evaporation Rate 55
803.3 Ambient Air Humidity, Enthalpy, and Humid Volume 56
803.4 Humidity Added by Fuel Combustion 56
803.5 Dryer Outlet Conditions 56
803.6 Dry Air Flow Rate 56
803.7 Total Air Flow Rates 56
803.8 Burner Heat Input 57
803.9 Check Driving Force for Evaporation 57
804 Example: Typical Dryer Test 59
804.1 Experimental Data 59
804.2 Humidities (H) 59
804.3 Air Flows – Outlet 59
804.4 Air Flows – Inlet 59
804.5 Dry Solids Balance 60
804.6 Moisture Balance – Evaporation Rates 61
804.7 Heat Balances: Heater Duty and Fuel Consumption 61
804.8 Heat Balance Over Dryer 62
804.9 Thermal Efficiency 63
804.10 Adiabatic Saturation 63
805 Test Data Sheet 64
806 Psychrometric Chart 65
807 Index 66
808 References 67
809 Tables and Figures 68

100 Purpose and Scope

101 Purpose

101.1 Purpose

The purpose of this procedure is to suggest a method and provide guidance for conducting and interpreting performance tests on spray dryers.

101.2 Summary

The text discusses the reasons for spray drying. These reasons are usually to produce a product with certain desired properties or with better efficiency than other methods. Full-sized spray dryers are usually tested to determine performance. Pilot dryers are tested to evaluate the properties of the product that can be produced and to choose suitable operating conditions. A systematic test program such as a factorial experiment plan may be carried out to determine the best conditions to achieve the desired results. A definitive test may be done on a spray dryer that has already been optimized for performance.

The text discusses how to plan the test in light of these objectives. It gives guidance on the variables affecting product properties and dryer performance, to decide which variables to evaluate. However, spray drying is a complex process. Multiple interacting variables have to be considered in the plan. In addition to using this procedure, it may be wise to consider obtaining the guidance of a person experienced in spray drying. Art is involved as well as science, especially in evaluating a new material or optimizing conditions for a plant spray dryer. Although this procedure will help to obtain data useful for design, it is not intended to explain how to design a spray dryer. Also, since the subject is complex, detailed study of this procedure and of other literature may be required in order to reach the objectives.

Typical spray dryer installations are briefly described. Checklists are given to aid in planning measurements and listing steps needed for a test. Safety must be considered in any test planning. Some of the important measurement methods are briefly described. Equations and calculation methods are given to carry out heat and material balances that can be used to check accuracy of test results.
Balances are given in three levels of detail (methods). The simplest method compares the heat lost by the drying air stream to the heat required to evaporate the water in the feed. The second method uses psychrometric charts to account for the energy and moisture in humid air and to make a graphical heat balance. The third method uses equations to consider all inputs to the heat and material balances. These methods are demonstrated by examples given in the Appendix. A sample data sheet is also provided.

In this procedure air is the drying gas and water is the liquid to be evaporated. Other noncondensable gases and liquids may be used. These usually present special safety requirements not covered in this procedure. The material and heat balances given can be adapted for the other systems.

101.3 Scope of Spray Drying

Spray drying is rarely cost competitive with two-step dewatering and solids-drying processes. The equipment is costly, and the addition of bag collectors for solids recovery from large volumes of exit gas may double the cost of a spray-dryer installation. These higher costs are usually justified on the basis of some improvement in product quality, such as particle form, size, product flavor, color, or heat stability. The major and most successful applications of spray dryers are for solutions, slurries, and pastes which (1) cannot be dewatered mechanically, (2) are heat-sensitive and cannot be exposed to high-temperature atmospheres for long periods, (3) contain ultra fine particles which will agglomerate and fuse unless they are dried in a dilute gas suspension, or (4) require the product to be in a powder form to fit the needs of its applications. In spite of these limitations, spray drying is widely used in industry because of the benefits it provides to many products.

101.4 Reasons for Testing Spray Dryers

The four types of tests usually performed on spray dryers are Exploratory, Performance, Acceptance and Troubleshooting. Exploratory tests are intended to determine whether a satisfactory product can be produced in a spray dryer, and under what conditions. The results may even be used to obtain data to design a spray dryer. Performance tests determine production capacity and/or thermal efficiency for an existing spray dryer while delivering an acceptable product. Acceptance tests determine whether a new spray dryer meets agreed-upon performance criteria. Performance tests that look into details of operation can be useful for troubleshooting drying problems.

Sections 300, 301, and 301.1 point out the need to pick the correct type of test (exploratory, performance, or acceptance), and how to develop a plan for each type of test. The test plan should support the business plan for the plant, i.e., whether to increase production, or to expand the range of products. Section 305 explains test plan preparation.

101.4.1 Commercial-Size Dryers

Typical reasons for conducting tests on commercial-size dryers are:

101.4.1.1 Performance Test

To check the capacity and performance capability of the dryer

- Determine the performance of a dryer under typical operating conditions.
- Determine the maximum dryer capacity under existing operating conditions.
- Determine the desirable operating range for routine control of the dryer.
- Determine the optimum operating conditions for product quality, cost effectiveness, fuel conservation, and minimum environmental impact.
- Compare the performance of a new dryer with design data.

Section 302.1 discusses types of tests to measure capacity. Section 302.2 discusses the importance of heat and material balances to validate the test measurements and points out the importance of steady-state operation. See also Section 501. Section 303.5 discusses the interaction of capacity and rate effects with product properties, an interaction that makes for complexity in planning and carrying out spray dryer tests. Therefore, it is necessary to define the required properties for your application before conducting the performance test. Also, the required operating conditions to achieve the desired properties should already have been determined by extensive exploratory testing before conducting a performance test.

101.4.1.2 Troubleshooting

It may help to run a dryer test when problems occur during production of a particular material. The detailed work done during the test may help to elucidate the cause of problems. Measurements will include operating conditions, calculations will include heat and material balances, and observations will include characteristics of the product, its behavior in the dryer, and behavior of peripheral equipment. The actual test work may be similar to that of a performance test, but the different objectives should be specified before the test. Securing the participation of outside experts may help in such a test.

Several sections of this procedure give information that may help during troubleshooting tests. They are:

301.4 Dryer Controls and Instrumentation

301.5 and 302.1.2 Peripheral Equipment

302.2 Heat and Material Balance

303 Variables Affecting Product Properties, especially 303.2 Atomization

400 Methods of Measurement and Sampling

101.4.1.3 Acceptance Test

Typically a dryer is tested during commissioning to assure that design rates can be met per contract. This usually involves not only confirming evaporation rate but also product rate, quality, and yield. This procedure is not intended to be used for guarantee testing. Instead, a plan for such a test should be developed by negotiation between buyer and seller. The test requirements may be similar to those for performance testing given above.

101.4.1.4 Exploratory Tests.

To define or test dryer for new products or design data

- Determine whether suitable product characteristics can be attained by spray drying.
- Gather data for design of new dryers of different capacity for similar products.
- Study specific dryer variables as they relate to product characteristics, e.g., particle size, particle shape, bulk density, dustiness, and heat sensitivity properties such as, taste, and odor.

Section 303 discusses how the dryer operating variables may be adjusted to achieve desired product properties. Table 303.6 lists variables and their effect on performance and on product properties. Section 304 gives a further list of variables that may affect the results, Section 502.1 gives some priorities in dealing with these variables. Section 301.7 suggests a systematic way to deal with multiple variables in the experiment plan.

101.4.2 Laboratory- or Pilot-Sized Dryers

Tests in commercial-size dryers will be costly. Often **exploratory tests** can be done in a smaller dryer, such as at the facilities of an equipment manufacturer or a contractor. The small size of these dryers allows rapid changeover of test conditions, and easy cleaning between tests. Also, a pilot dryer test is unencumbered by any need to produce salable material, or to meet a production schedule, and may be operated by research personnel rather than plant personnel.

101.4.3 Scale-Up Considerations

Some test conditions available in commercial dryers cannot be achieved in a pilot dryer, because pilot dryers may not have the geometry to successfully dry all products. An evaluation must be made to take into consideration the limitations of a small dryer, particularly the limited residence time and the small diameter. The spray droplets may hit the wall and stick before they evaporate if the drying temperature is low. Especially, it may not be feasible to use a disc atomizer, which tends to throw the droplets toward the wall. Therefore pneumatic spray nozzles are normally used in pilot dryers, whereas disc atomizers are often used in commercial dryers because of their greater throughput. They may produce a different particle size distribution than nozzles. Furthermore, to achieve a dry product in a pilot dryer it may be necessary to operate at a higher temperature than desired, causing damage to some properties such as taste, odor, and color. Still, the test may be useful if it provides samples of product that indicate other properties such as form, density, and flowability, and thus gives clues for further testing on a larger scale.

Although a heat balance can be made on a pilot dryer, the results may not be applicable to a scaled-up dryer. Different surface-to-volume ratio of commercial equipment will result in different heat balance and energy efficiency conditions. The heater itself may be quite different. Thermal insulation and surroundings of the dryer may differ. Some of these effects can be corrected for by engineering calculations. However, measurement of thermal efficiency in a pilot dryer can give only a rough guide to efficiency in larger-scale production, but heat balance and thermal efficiency measurements will be important in large-scale dryer testing.

Some variables that were not thought of during pilot testing may arise during testing on commercial-scale dryers, such as different feed mixing techniques used by different plant operators. The product collection equipment may be different, but this should not affect the dryer.

102 Design vs. Operational Variables

A program of dryer testing can lead to optimum operating conditions for the dryer tested. In addition to the obvious effect on capacity, design can affect product properties. For example, chamber shape can affect residence time distribution, which in turn affects exposure of a product sensitive to heat. Thus the importance and limitations of dryer design should be recognized in an evaluation made prior to the actual test.

103 Liability

AIChE and members of the various committees involved make no representation, warranties or guarantees, expressed or implied, as to the application or fitness of the testing procedures suggested herein for any specific purpose or use. Company affiliations are shown for information only and do not imply procedure approval by the companies listed. The user ultimately must make his own judgment on the specific testing procedures he wishes to utilize for specific applications.

200 Definitions and Descriptions of Terms

201 Dryer Design

201.1 Dryer Chamber

A spray dryer consists of a stationary vessel or chamber. It can have several shapes (Fig 201) and includes an atomizer to disperse the feed into the gas stream. The gas inlet to the dryer can be located at the top, bottom, or side, depending on the configuration. There is usually some method of gas distribution -- perforated plates, louvers, or vane rings -- within the chamber. The gas outlet will usually be at the opposite end of the chamber from which feed is introduced. If the gas outlet is the only outlet of the vessel, the spray dryer product is carried in the outlet drying gas. The vessel may be equipped for separation of the product by gravity, in which case, the chamber will have two outlets.

201.2 Airflow Patterns

Airflow patterns depend on chamber design and define gross airflow (as distinguished from localized eddies and backflow). (See Fig. 201) Parallel flow spray dryers usually tall and narrow to minimize back mixing of gas, and to expose sensitive dry product to only the cooler exit air, as in Figure 202.

Counter-current flow dryers are used to expose the product to hot air. While this design removes tightly bound residual water more efficiently, and it also results in less expansion of the product, it is not feasible for sensitive materials.

Atomization and evaporation create complex localized flow patterns around the atomizer. These patterns usually contain some circular flow components. In mixed flow dryers, a portion of the air is supplied tangentially. Also, cooling air may be introduced at the exit, to protect a sensitive product.

201.3 Product Flow

Depending on the dryer design, product flow can be co-current or counter-current to gas flow. The dryer can be designed to either aid or counter particle separation by gravity or centrifugal forces. In this way, the designer can alter the residence time distribution of the larger and smaller particles, to aid in drying the larger particles or exposing particles longer to drying conditions.

201.4 Atomizer

The atomizer may be a high-speed rotating disc, a two-fluid (pneumatic) nozzle, or a single-fluid (pressure) nozzle. The function of the atomizer is to break the feed material into many droplets to increase the surface area and to form the droplets that will result in the desired particle size and shape for the final product. See 303.2.

201.5 Heating Methods

The gas stream may be heated directly or indirectly by any convenient means before entering the dryer cylinder. This gas stream provides the thermal energy needed to:

1. Heat the material to be dried.
2. Heat and evaporate the liquid and heat the vapor to the exhaust temperature.
3. Compensate for conduction, convection, and radiation heat losses from the chamber.
4. Provide the energy required for the heat of reaction, if applicable.
5. Provide the energy required for the heat of crystallization, if applicable
6. Avoid conditions where vapor might condense in the exit stream. See Section 604.2.3

201.6 Product Recovery

After leaving the drying chamber, all gases and vapor from drying usually pass through product/dust recovery equipment before being released to atmosphere. Cyclonic separators for product recovery, followed by dry fabric filters or wet scrubbers for dust collection, are the most common forms of separation equipment employed on spray dryers.

Figure 201. Typical Spray Dryer Configuration

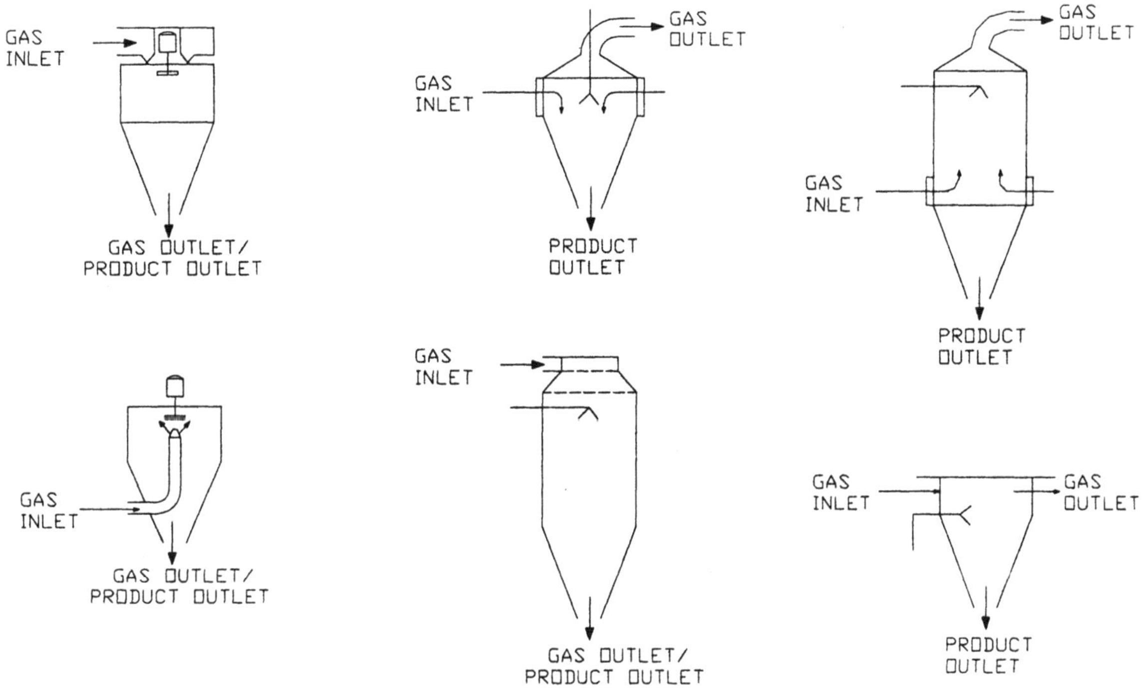

Figure 202. Exploded View of a Spray Dryer System

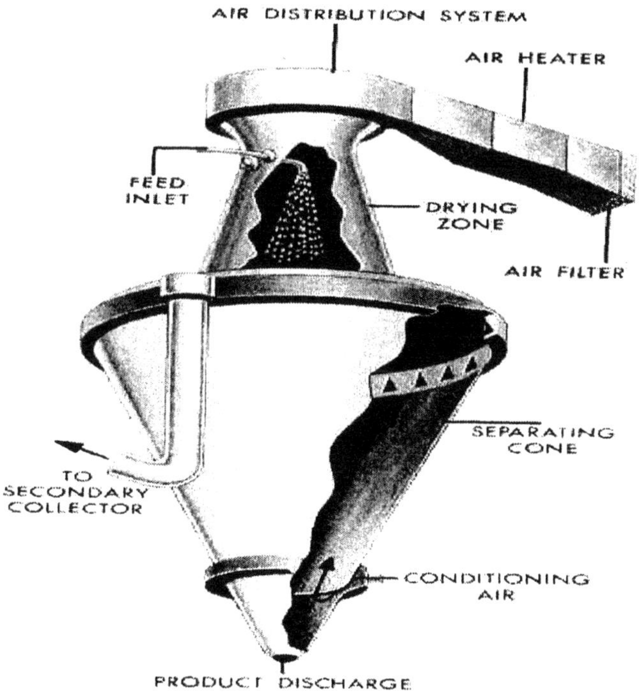

Operating diagram of a parallel-flow
SWENSON spray dryer for special products

201.7 Airflow Motive Force

Fans induce the gas flow through the heater, drying chamber, and product/dust collection equipment. The most simple fan arrangement, which is used on dryers with low pressure drop heaters, is a single induced-draft exhaust fan located downstream of the dust collection system so the fan will operate on clean gas. This one fan, therefore, must have static pressure capability to accommodate the pressure-drop through the inlet gas filter, gas heater, gas inlet duct, and dust collector, which typically amounts to 3,750 to 12,500 Pa (15-50 inches of water).

Alternatively, when the air leakage into the dryer system seems excessive, the second forced-draft fan may be installed upstream of the heater inlet to overcome the inlet system pressure drop. By balancing the two fans in this manner, the pressure inside the spray-drying chamber can be adjusted to a level within 125 Pa (0.5 inches of water) of atmospheric pressure to minimize leakage to or from ambient air.

202 Description of Terms

202.1 Drying

Drying is an operation that separates a liquid from a solid or semisolid material by evaporation. Dehydration refers to the drying of vegetable and animal products to less than their natural moisture content and removal of water-of-crystallization from chemical compounds. Following is the terminology commonly employed for various drying and dehydration processes.

- <u>Absolute Humidity</u> is the amount of liquid (e.g., water) vapor in a given gas stream, expressed as weight of liquid per weight of dry gas (in kg/kg).

- <u>Atomization</u> is the formation of a spray by breaking up bulk liquid plus solids (dissolved or suspended) into many individual droplets. Its principal effect is to produce a high ratio of surface to mass in the liquid phase, which promotes drying and controls particle size of solid products.

- <u>Constant-Rate Period</u> is an initial period during which enough liquid reaches or remains on the surface of particles to maintain a constant rate of evaporation.

- <u>Density</u> is the mass per unit volume of material, kg/m^3. It may be measured in several ways. <u>Bulk density</u> is measured by weighing a known volume in the as poured, or loose condition. A <u>settled, tapped, or packed density</u> is found by tapping the container. <u>Particle density</u> is that of particles without interstitial volume between them. <u>Absolute density</u> is that of material with no voids. <u>Particle density</u> is different from <u>Absolute density</u> because a particle may be a hollow sphere and thus the particle occupies a defined volume but has much less mass than a solid particle of the same dimension.

- <u>Dew Point</u> or saturation temperature is the temperature at which a given mixture of liquid (*e.g.,* water) vapor and gas (air) is saturated. At the saturation temperature, the liquid exerts a vapor pressure equal to the partial pressure of the liquid vapor in the given mixture. Below this temperature, liquid will condense. This can result in mist or condensation on cold surfaces or product.

- Diffusion-Dominated Drying is drying during the falling-rate period, when diffusion of moisture from within the particle limits the drying rate, rather than heat transfer from the gas to the surface of the particle.

- Direct Drying occurs when the heating medium, usually hot combustion products, comes in direct contact with the particles. It is said to be adiabatic because when the dryer is insulated there is no gain or loss of energy by the system.

- Direct-Heat Dryer is one of a class of drying equipment in which heat is transferred to the atomized droplets being dried by direct contact with the heating medium. The heating medium is usually a hot gas.

- Dry Basis expresses the moisture content of a wet material as a fraction of the weight of dry material. The advantage of using this basis is that the difference between the moisture content before and after drying is the moisture change per unit weight of dry material.

- Dryer Efficiency is the fraction of the thermal energy supplied that is used to heat and vaporize the liquid and to heat the vapor to the dryer exhaust temperature.

- Dust Loading is the amount of powder in a gas stream, kg/m^3 (mg/m^3).

- Enthalpy is the heat content of a material, kJ/kg. For a mixture, it is the weighted sum of the individual enthalpies. Enthalpy is measured above a base temperature, 273 K (0 EC).

- Evaporative Efficiency of the gas stream flowing through a direct-heat dryer compares the amount of evaporation actually obtained to that which could be obtained if the gas stream were saturated adiabatically before leaving the dryer.

- Falling-rate Period is a drying period during which the drying rate per unit of material surface continually decreases, generally after surface moisture has been removed. The liquid or vapor has to diffuse from within the particle.

- Heat of Vaporization is the change in enthalpy involved in the transition from liquid to vapor, expressed as kJ/kg (kcal/kg).

- Humid Heat is the combined sensible heat of the air stream and the heat of vaporization of its water vapor content. (See Section 602.3.2)

- Indirect-heat Dryer is one of a class of drying equipment in which the heating medium is physically separated from the material being dried by a barrier. The heater heats the inlet gas primarily by conduction and radiation through the barrier, and the hot gas heats the feed by conduction.

- Internal Diffusion occurs in a material during drying when the liquid evaporation rate appears to be controlled by diffusion of liquid or vapor flow to the particle surface.

As the particle dries the diffusion path lengths and causes reduced mass transfer. This results in the falling rate period.

- Latent Heat is the energy involved in a phase change (e.g., liquid to gas) that does not result in a temperature change expressed as kJ/kg (kcal/kg).

- Moisture Content of a material is the moisture quantity per unit weight of dry or wet solid. The basis should be specified. See Dry Basis, above.

- Moisture Gradient refers to the moisture profile in a material at a given moment during a drying process. The nature of the moisture gradient depends on the mechanism of moisture flow inside the material, *i.e.* internal diffusion. For fine particle products, it cannot be measured, but it can be calculated from the moisture content and the diffusivity of water in the solid.

- Percent Saturation of a gas containing a condensable vapor is the ratio of the partial pressure of the condensable vapor in the gas to the vapor pressure of the pure condensable material as a liquid at the same temperature, expressed as a percentage. For water in air this is also called percent relative humidity.

- Residence Time Distribution of the particles is spread over the mean residence time of the gas because of mixing in the chamber, and settling.

- Residence Time – Mean residence time of the gas is the ratio of the chamber volume to the volumetric flow rate. It may differ from the residence time of the solids.

- Sensible Heat is the energy involved in changing the temperature of a substance expressed as kJ/kg °K (kcal/kg °K).

- Swirl Chamber is used in pressure nozzles that produce a hollow cone spray pattern. The fluid enters the chamber tangentially and acquires centrifugal motion, which flings it outward as it exits the orifice. Vanes can also achieve this effect.

- Thermal Efficiency of a dryer is defined (*Reference 808.1, p 111*) as the ratio: Heat used in evaporation to Heat input.

- Wet Basis expresses the moisture content of a wet material as the weight of moisture per unit weight of dry material plus the contained moisture.

- Wet Bulb Temperature is the dynamic equilibrium temperature attained by a liquid surface when the rate of heat transfer provided to the surface by convection from the gas phase equals the latent heat taken up by the evaporation or mass transfer away from the surface.

300 Test Planning

301 Preliminary Objectives

The cost of performing a plant test goes far beyond the time and material expended during the actual test run. Careful planning and preparation are essential to maximize the economic and technical benefits of a test. A poorly run test may result in the loss of a business opportunity.

301.1 Test Objectives

301.1.1 Inter-related Objectives

Usually the most important objectives are to determine capacity, energy efficiency, and product properties. These objectives are inter-related. Section 303 includes a discussion of how the operating parameters may be varied to achieve desired properties.

301.1.2 Specific Objectives

Specific test objectives should be stated in writing. This phase of test planning should include techniques for determining whether or not the objectives are being met while the test is in progress.

301.1.3 Capacity vs. Design

A new dryer can be designed to achieve practically any desired production rate while delivering the desired product qualities. The design is based upon an evaporation rate or capacity at specified operating conditions derived from a required product rate at specific inlet and outlet concentrations of moisture (a design material balance).

In contrast, the production rate of an existing dryer will be limited by its design. For example, the manufacturer sets the inlet airflow rate. In dealing with an existing dryer, one must usually be willing to accept a production rate achievable under conditions that give the desired product qualities. The objective of testing an existing (full size) dryer will usually be to determine these conditions, and secondarily to determine achievable production rate under these conditions.

301.2 Organizational Resources

The test objectives help determine which organizational resources should be involved in the test. Plant tests frequently involve people from many disciplines such as plant maintenance, analytical services, design organization representatives, and equipment supplier representatives.

301.3 Schedule

It is (obviously) best to schedule a plant test for a time when the plant is expected to have time available on equipment. However, many other variables must be considered for a successful plant test such as availability of plant feed stocks and raw materials, conditions of upstream and downstream units, the time it will take to complete the required preparations, personnel availability, availability of analytical services, and the anticipated ambient conditions. If the plant test must be conducted under extremely unfavorable conditions, alternative plans may be necessary. The length of test will depend on the test objectives, the number of conditions tested, and the required time to run at each test condition.

301.4 Dryer Controls and Instrumentation

The test participants should understand the spray dryer system control scheme so they can anticipate the equipment's response to deliberate or consequential disturbances. Equipment diagrams should be available showing the locations for temperature, pressure, flow rate, and humidity measurements and sample points. Accuracy provided by the dryer instrumentation is an important factor for a successful test. It may be necessary to recalibrate critical instruments and tune controllers prior to starting the test. It may also be necessary to make preparations for, and carry out all piping, instrument and equipment changes where necessary to obtain the operating conditions, samples, and data required. Be certain of the information that the instrument is providing (i.e. is the flow rate temperature or pressure compensated, what are the units of measure). Record external environmental conditions during the tests, *i.e.,* barometric pressure, air temperature, and air humidity. Annual averages for these conditions for the site should be available to aid in the proper interpretation of the results and prediction of full-scale production. Convenient laboratory facilities should be made available to check the moisture content of the feed and product and microscopic and screen evaluation of particle size and shape.

301.5 Peripheral Equipment

The potential limitations of peripheral equipment such as pump, preheater, fans, valves, cyclones, and bag houses should be considered. Assure that a good supply of the desired feed is available.

301.6 Pretest Calculations

Pretest calculations of the test conditions can be useful to check against the capacity of fans, heaters, dryer, and product recovery. Also check the requirements for separately storing the test product grades from each test, since they may be nonstandard and variable material or quality. Review the environmental limitations on emissions and operating conditions for their effect on the test conditions. For orientation, it may be well to study the examples in Section 800.

301.7 Test Plan

Although the factors affecting spray dryer performance can be very complicated, there are a few variables that can be adjusted to help control final product characteristics. General trends are described below. Exceptions can arise, and other factors can affect, hide, or hinder the desired or expected results. Actual product tests are often needed to confirm and evaluate the ability to control and produce desired characteristics in a given product.

301.7.1 Handling Multiple Variables

So many variables affect spray dryer performance that a partial factorial experiment plan may be required to sort out the effect of these variables. A Factorial Experiment Plan is a plan for carrying out experiments laid out in a multidimensional matrix, usually at upper and lower conditions for each variable.

The matrix is symmetrical so that all variables are equally represented. In a Partial Factorial Experiment Plan only a fraction of the possible combinations of variables are included, but all are represented at some high and low values. See *Reference 808.7, Section 808, and Reference 808.3, pp 3-87 to 3-88.*

Initially, sufficient knowledge may not be available to choose these conditions. Under these conditions a preliminary Evolutionary Test Plan may be best. Start with any set of conditions chosen either from literature sources (such as Table 301 and Table 302 or *Reference 808.3, pp 12-81 to 12-90*) or recommendations from equipment manufacturer or others, or simply choose conditions that are convenient. Conduct a series of tests, varying first the variables expected to have the most effect on the desired properties. See Section 303 to determine these. For heat sensitive material, see Sections 303.3 and 502.

TABLE 301. Performance Data of Spray Dryers

Reference 808.6, p 275

Kind of Stock	Moisture Content In (%)	Moisture Content Out (%)	Spray Device	Flow Pattern	Air Temperature In (°C)	Air Temperature Out (°C)
Skim Milk, $d = 60 \mu m$	48-55 50-60	4 4	Wheel or Nozzle 170 to 200 bar	Parallel Parallel	250 250	95-100 95-100
Whole Milk	50-80	2.5	Wheel or Nozzle 100 to 140 bar	Parallel	170-200	
Eggs, whole	74-76	2-4	Wheel or Nozzle	Parallel	140-200	50-80
Eggs, Yolk	50-55	2-4	Wheel or Nozzle	Parallel	140-200	50-80
Egg. Whites	87-90	7-9	Wheel or Nozzle	Parallel	140-200	50-80
Coffee, Instant, 300 μm	75-85	3-3.5	Nozzle	Parallel	270	110
Tea, Instant	60	2	Nozzle, 27 bar	Parallel	190-250	
Tomatoes	65-75	3-3.5	Wheel	Parallel	140-150	
Food Yeast	76-78	8	Wheel	Parallel	300-350	100
Tannin	50-55	4	Wheel	Parallel	250	90
PVC Emulsion, 90%>90 μm <60 μm	40-70	0.01-0.1	Wheel or Nozzle or Pneumatic	Parallel	165-300	
Melamine-urethane-formaldehyde resins	30-50	0	Wheel, 140-150 m/sec	Parallel	200-275	65-75
Heavy duty detergents	35-50	6-13	Nozzle, 30-60 bar	Counter	350-400	90-110
Kaolin	35-40	1	Wheel	Parallel	600	120

TABLE 302. Performance of a Dryer 18 ft Diameter by 18 ft High with a Spray Wheel and Airflow of 11,000 cfm at the Outlet

Reference: 808.6, p 275

Material	Air Temp (°K)		% Water	Evaporation Rate
	In	Out	In Feed	(kg/s)
Blood, Animal	440	345	65	5.9
Yeast	500	335	86	8.2
Zinc Sulfate	600	380	55	10.0
Lignin	475	365	63	6.9
Aluminum Hydroxide	590	325	93	19.4
Silica Gel	590	350	95	16.9
Magnesium Carbonate	590	320	92	18.2
Tanning Extract	440	340	46	5.2
Coffee Extract	420	355	70	3.8
Detergent A	505	395	50	5.0
Detergent B	510	390	63	6.2
Detergent C	505	395	40	2.6
Manganese Sulfate	500	415	50	5.5
Aluminum Sulfate	415	350	70	1.7
Urea resin A	535	355	60	3.8
Urea Resin B	505	360	70	1.9
Sodium Sulfide	500	340	50	2.0
Pigment	515	335	73	13.2

301.7.2 Safety

Any equipment testing must conform to the latest requirements of applicable safety standards. These include, but are not limited to plant, industry, local, and state and federal regulations. It is recommended that all testing be conducted under the supervision of personnel fully experienced in plant and equipment operating practices. A safety review should be conducted prior to operating any dryer outside of the original design parameters.

Most spray drying systems are heated by natural gas or propane burner systems. Only qualified personnel familiar with the related safety regulations should adjust or modify the burner safety control systems on these heaters.

Explosions are a particular hazard involving dispersions of organic powder in air. A range of explosive concentrations and temperatures exists for most combustible powders in air; these properties can be found in the material's MSDS. A spray dryer may contain an air ignition source. Explosions may also occur in peripheral powder handling situations. The operators should be aware of this possibility and guard against it. For example, see Section 402.1. Certain dusts also expose workers to touch and/or inhalation hazards.

A combustible powder should not be dried in a dryer that has not been designed for the hazards of that particular material. Older spray dryers were often not built to current standards for materials that are dust explosion hazards. This is also true for flammable solvents.

301.8 Environmental

The test procedure must conform to the latest requirements of all applicable environmental standards, which include plant, industry, local, state and federal regulations. Environmental conditions that apply to the equipment in normal operation should also apply during testing.

301.9 Cleaning and Inspection

Before carrying out any test, open the dryer chamber and inspect the internals for damage, misalignment of parts such as nozzles, and for presence of material from previous tests on walls or nozzles. A light dusting of material may be removed by brushing, but caked materials must be washed off. Inspection may reveal problems with operating conditions of the previous test and help guide the next test.

301.10 "Dry Run"

A useful procedure is that of conducting a "dry run" before the official test. The pretest should include a walk-thru at the equipment following all of the actions of the operators, technicians, analyses, data collection, and data calculations. This will enable identification of shortcomings in data collection, sampling points, sample handling, analytical and sequencing of conditions. Check the blower(s) and feed pump(s) for lubrication and proper rotation. Calibrate all instruments, including air and liquid feed flow instruments.

302 Types of Test

The types of test that are usually run are summarized in Section 101.2. These fall into three general categories; Capacity, Product Properties, and Acceptance Test. All are discussed below.

302.1 Dryer System Capacity

Tests can be performed on new or old dryers to determine how the dryer is operating, whether it can be adjusted to achieve more capacity, or what capacity may be feasible for new operating conditions to achieve desired properties of new products. Some large spray dryers may have capacity for evaporation rates up to 6,800 kg/hr (15,000 lb/hr) or more, but powder production rate depends on many variables.

The dryer system capacity may be controlled by the limitations of the drying chamber itself, or by limitations of peripheral equipment such as the feed system, the atomizer, fan capacities, energy input, product recovery equipment, and environmental controls. Comparison of mass and energy balances with equipment capacities will help analyze which part of the system is likely to restrict capacity. (Also, see Section 502.1.)

302.1.1 Dryer Chamber Limitations

There is a limit to how much capacity can be increased by changing peripheral equipment, such as increasing heat input and fan size for a given dryer. The design of the drying chamber itself limits capacity because of the interaction of the equipment with the product properties. Increasing the severity of drying conditions will eventually affect product properties. Product properties include required moisture content, flow properties and other quality measures. Details of these product property interactions are discussed in Section 303.

302.1.2 Peripheral Equipment Limitations

Initial design of the dryer should have properly fit or balanced all components to provide good drying conditions. In some cases design or maladjustment of peripherals may limit the performance. However, a specific application may deviate from assumptions made in the initial design. Therefore, it may be important to separately test the peripheral dryer components and evaluate their design. If the chamber design will support increased capacity, it may be desirable to investigate whether changes to the feed system, fan capacities, energy input, product recovery, and environmental controls will result in increased capacity.

Analysis of peripherals could include determining and confirming:

- The energy rating for the air heater. Can it reach its original design?
- The airflow rating for the system fans. How much can they be varied without causing problems, such as unsatisfactory chamber residence time or drying time?
- Duct size restrictions, transport velocity, dust collector or scrubber performance limitations. Will they cause airflow limitations?
- Feed pump flow rate and pressure- range and limits. Do they fit the test ranges and atomizer requirements?
- Atomization limits: Nozzle number and size, , rotary atomizer limits (volume or horsepower). Are they compatible with the test conditions?

In addition, confirming the capacity and performance of each component can help find and determine if losses are preventing intrinsic capacities from being reached (i.e. air leaks, poor insulation, poor component adjustment or maintenance.)

302.2 Heat And Material Balance

Another objective of testing may be to perform a heat and material balance to assure that the dryer is performing satisfactorily. In most cases, one can do a reasonable check on performance just by measuring production and heat input. However, instructions for a detailed heat and material balance are also included in Sections 603 & 800s. For this part of the test, it is important to determine that dryer operation is steady, and that the feed and product are adequately sampled and characterized.

302.2.1 Heat Balances

The major heat load Q for the dryer is the evaporation requirement, determined by the required production rate. In theory ($Q = L\ Cp\ \Delta T$) the same heat load Q can be provided by less process gas L at higher temperature drops (ΔT), or more process gas L at lower temperature drops (ΔT). The dryer is designed to provide this capacity at operating conditions (L, ΔT) to give good product. The design includes the component requirements: heater, fan, atomization technique, and chamber design and size. (See Computation, Section 600.)

302.2.2 Economic Optimum Production Rate (Thermal Efficiency)

Optimum conditions for thermal efficiency use the highest ΔT, i.e., the highest inlet gas temperature and lowest outlet temperature that will produce good product. Increasing inlet gas temperature increases evaporative capacity of the dryer, but also is likely to affect product properties, such as bulk density due to formation of more hollow particles, and may increase thermal degradation. Therefore, trial-and-error testing is required to determine this optimum. Section 303 discusses the effect of operating conditions on properties as a guide for testing.

302.2.3 Steady State Requirements

The performance capability of a spray dryer can be demonstrated only under conditions of steady-state flow of material and gas. Also, the feed material moisture content and temperature, the gas temperature and humidities into and out of the dryer, and the product moisture content and temperature must remain essentially constant during the test period.

302.2.3.1 Major Measurements

During the test, measurements must be taken of gas and material temperatures, moisture contents and flow rates, total heat input to the dryer, and heat losses from the drying chamber. It is necessary to record the quantity, temperature, and moisture content of material going to product collection equipment separately from the chamber product. This is necessary because the temperature of material conveyed to the product recovery equipment can be higher than the temperature of the product from the dryer, but it will not be higher than the gas temperature. There also may be a difference in residence time of the hot gas. Without an accounting of material division, a material balance on moisture content and an accurate accounting of heat consumed as sensible heat in the material is not possible.

302.2.3.2 Product Recovery Equipment

Apart from the need to close heat and material balances around the dryer, it is not essential to carry out a performance test on product recovery equipment at the same time as the dryer performance test.

302.2.3.3 Consistency Test

Most measurements made on commercial-sized dryers are susceptible to human, instrument, and analytical errors. In addition, small but uncontrollable variations usually occur in material flow, gas flow, temperature and moisture contents during the performance test. Therefore, the heat and material balance results must be checked to be verify that they are consistent among themselves. The balances to be obtained must be as follows: (See Section 600 and 800)

302.2.3.4 Moisture

(Feed moisture content) – (product moisture content) – (dust moisture content) = (Evaporation).

(Evaporation) + (moisture from direct-fired fuel combustion) = (gas stream humidity gain).

302.2.3.5 Product

(dry material flow in) = (dry material flow out).

302.2.3.6 Gas

(dry gas flow in) = (dry gas flow out)

302.2.3.7 Energy

(heat gained by the gas through the heater) = (heat provided by fuel burner or other heat source)
(heat lost by gas through the dryer) = (material sensible heat gain) + (vapor sensible heat gain) + (heat of evaporation) + (dryer heat losses) + (heat of reaction/crystallization).

302.2.4 Data to Be Collected

In order to complete these balances, the appropriate data listed and arranged in groups in Section 304 should be collected.

302.2.5 Alternate Fuel Measurement Methods

The measured fuel consumption of a direct-fired, direct-heated dryer should be compared with the calculated fuel consumption by chromatographic or equivalent analysis of the exit gas. If this method is used, a complete fuel analysis is required and the effects of air leakage must be considered.

302.2.6 Gas Flow Measurement by Cyclone Pressure Drop

Product collection cyclone pressure-drop characteristics usually are available from the manufacturer. Pressure drop should be expressed as a number of gas velocity pressure heads based on the inlet gas velocity and density. With this information, the pressure drop can be used as an additional measurement of gas flow to compare to other flow measurements.

302.2.7 Fan Performance Data

The fan manufacturer's performance data and fan curves showing volume flow versus static pressure versus power consumed at the operating fan speed can be used to estimate air flow. In the United States, published data are usually based on air at 294°K (70°F) and these data often must be corrected for actual fan gas density. On the other hand, most manufacturers will provide fan curves corrected for specific applications, if asked to do so. Actual drive motor current and voltage along with motor efficiency data will be required to determine actual power input to the blower.

302.2.8 *Inlet Gas Flow Measurements*

To measure inlet gas (air) flow on dryers provided with direct-combustion air heaters measurements of temperature, pressure, humidity and flow of both primary air supplied for fuel combustion and the dilution air supplied for dryer inlet gas temperature control will be required.

302.2.9 *Heater Coil Pressure Drop Data*

Inlet gas (air) flow to dryers provided with steam-coil air heaters can be estimated by measuring the static pressure drop of the gas flowing through the heaters. Most steam-coil manufactures provide pressure drop data for their specific coils based on gas velocity, gas density, number of coil rows deep, and fin density. Sometimes coil pressure drop measurements can be substituted for a direct measurement of inlet gas flow. However, such a measurement is usually recommended as a way to confirm a direct gas flow measurement and, perhaps, uncover other measurement errors.

302.3 Product Properties

The product properties usually of the greatest interest are

- Water Content
- Particle Size
- Particle Shape
- Flow Properties
- Bulk density
- Dust content

The process conditions affecting these properties are discussed in Section 303.

302.4 Acceptance Test

An acceptance test is a special type of performance test. Ideally, the conditions under which it is run are those for which the equipment was designed or guaranteed. The purpose of the information is to check that the capacity and product quality are as guaranteed.

302.4.1 Extrapolated Test Results

Those responsible for operating the unit should make every effort to achieve design feed rates and compositions for a reliable acceptance test. Frequently, however, the conditions for the test can not be the same in all respects as those for which the equipment was designed. In this case, a test can still be made to check the dryer system, and test results can be extrapolated to the design conditions by techniques based on sound engineering principles.

303 Variables Affecting Product Properties

303.1 Residual Moisture Content

Increasing outlet temperature reduces the gas relative humidity and increases the drying rate, thus producing a lower residual moisture product. The inlet gas temperature can be raised to raise outlet gas temperature, unless the inlet gas temperature is already at the maximum that will not degrade the product. Alternatively, the feed rate can be reduced to raise outlet temperature, at the cost of decreased production capacity. It is more efficient to reduce the moisture content of the feed, as discussed in Section 604.2.1

Reducing the airflow allows a longer residence time of particles/droplets, and hence a longer drying time. However, this will have a small effect on moisture content. Reduced airflow will also reduce the heat input and hence the drying capacity. Changing airflow will also influence the product collection efficiency and separation depending on the system arrangement. These changes may include separation in the drying chamber, in cyclones, or dust collector, etc. Reducing droplet size and hence particle size will lower moisture content of product, especially with materials having tightly-bound water.

Lower residual moisture content can be achieved by further drying the product in other equipment, such as a fluid bed dryer. This may be necessary if the residual moisture is tightly bound. For some materials, heat sensitivity is less after the moisture content has been reduced. Thermal degradation may not occur in this subsequent drying step, even though the drying time is long.

303.2 Atomization

Proper atomization of the feed is the key to successful spray drying. One of the following atomizing devices is usually used: (1) pressure nozzles, (2) two-fluid (pneumatic) nozzles, and (3) centrifugal disks. With these atomizers, thin solutions and slurries may be dispersed into droplets as small as 2 μm. The largest drop sizes rarely exceed 500 μm (35 mesh). Atomizer performance ranges are given in Table 303.4.

Table 303.4 Particle Diameters, Densities, and Energy Requirements

Reference 808.6, *p 276*

(a) Atomizer performance

Type	Size Range (μm)	Power Input (kwh/m^3)
Single fluid nozzle	8 - 800	0.3 – 0.5
Pneumatic nozzle	3 – 250	--
Spray wheel	2 – 550	0.8 – 1.0
Rotating cup	25 - 950	--

(b) Dry Product Size Ranges

Product	μm
Skim milk	20 – 250
Coffee	50 – 600
Eggs	5 – 500
Egg white	1 – 40
Color pigments	1 – 50
Detergents	20 – 2000
Ceramics	15 - 500

303.2.1 Nozzles

The primary variables for one fluid nozzles are the orifice diameter and swirl chamber diameters. See Figure 300. Increasing orifice size substantially increases feed rate for a given pressure and increases droplet size. Decreasing swirl chamber diameter results in a wider hollow cone pattern improving dispersion of the droplets into the drying medium. The second variable is applied pressure. For one-fluid nozzles, the pressure affects the particle size and size distribution as well as the feed rate. For two-fluid (pneumatic) nozzles, the gas pressure affects the droplet size, while liquid pressure affects the feed rate. These two variables can be separately adjusted (although there is some interaction).

Two-fluid nozzles are usually not economical for large-scale spray drying, because of

1. The cost of large volumes of pressurized air required,
2. The very low nozzle efficiency results in an installation with a large number of expensive nozzles requiring maintenance,
3. The reduction of spray dryer capacity due to the cooling effect of atomizing air entering at the hot end of the dryer. *(Reference 808.1. p.263)*

Two-fluid nozzles are often used in laboratory- or pilot-scale tests.

Figure 300. One Fluid Spray Nozzle with Swirl Chamber

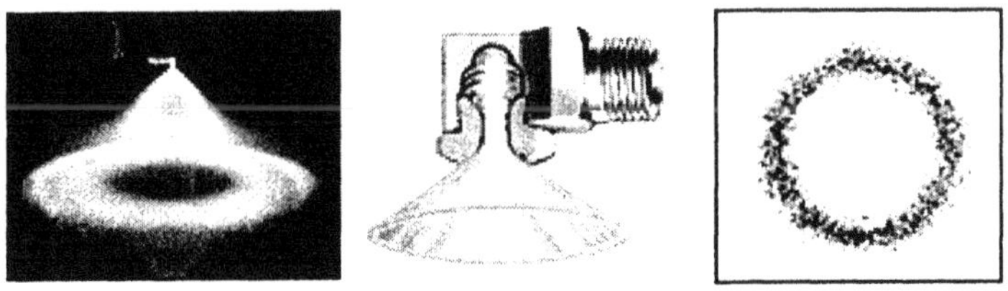

303.2.1.1 Nozzle Performance

Increasing feed rate increases the droplet size for both types of nozzles. Flow rate and droplet size in one-fluid nozzles are affected by fluid pressure, but individual nozzles have a rather narrow range of operation. This means that selection of the proper nozzle size and design is important. The subject is outside of the scope of this procedure, but Masters (*Reference 808.1, Chapter 6*) gives extensive information. Also, nozzle manufacturers give advice and performance data on their nozzles. However, the performance on a solution or suspension will differ significantly from performance on water, so charts from brochures should not be relied on. Instead, pre-test measurements should be made to determine the nozzle flow rate at several pressures, to guide selection of operating condition during the spray dryer tests. The resulting particle size distribution can best be determined by actual spray dryer tests.

A spinning-disk atomizer has a greater capacity than a nozzle, yet for a large spray dryer a bank of 20 nozzles can handle an even greater flow with more uniform distribution of spray. The flow pattern of the spray is also different, and needs to fit the air flow characteristics of the dryer chamber. The spray from nozzles is directed downward, while that from disks is outward across the airflow. Effect of pressure is important with nozzles, and pressures up to 1000 psig or more are sometimes used.

303.2.1.2 Nozzle Maintenance

Attention should be given to nozzle maintenance (*Reference 808.1, p. 245.*) To maintain nozzle performance, periodic cleaning is essential, and parts should be handled carefully to prevent damage. Nozzle parts are accurately machined, and variations in dimensions between individual parts can also affect results. Dirt, clogging, damage, wear or improper nozzle assembly can result in spray asymmetry, with wide fluctuations in feed rate, and a product that does not meet size specifications. It is standard practice to dismantle the nozzles during shutdowns, remove any deposits with a stiff brush, wash all parts, and store under water before use.

303.2.1.3 High Pressure Pumps

Single fluid pressure nozzles must be fed with a high-pressure pump in order to achieve proper atomization. Typical pressures up to 3,000 to 4,000 psi and occasionally up to 5,000 psi (20,700 kP to 34,500 kP) are usually provided by triplex piston or plunger pumps. These pumps generally require an inlet pressure of 30 to 50 psig (207 kP to 345 kP) to prevent cavitation and subsequent pump damage. Maintenance of these pumps is critical to maintain flow and pressure control. Inlet and outlet check valves, ceramic cylinder liners and pistons as well as the pump seals are all areas that can experience pluggage or wear. Installation of an inline spare can ensure a high on-stream factor.

303.2.2 Spinning Disks

Spinning-disk atomizers offer several advantages over nozzles (*Reference 808.1, p. 221*), especially for large spray dryers, even though they require precision manufacture and cost more than nozzles. They produce highly homogeneous sprays over a wide range of mean droplet sizes. Varying the disk speed, without changing the design can readily control droplet size. There is little chance of clogging. The wide projection of spray into the drying air gives effective spray-air contact resulting in fast drying rates. Unfortunately, this means that droplets will strike the dryer walls if the dryer diameter is small: (as in a pilot dryer) Manufacturers offer technical help in selecting atomizers and in determining proper operating conditions.

303.2.2.1 Disk Performance

The speed, diameter and design of a spinning disk affect the droplet size. Also, increasing feed rate increases the droplet size. Consult with the atomizer vendor on the effects of different disk design.

303.2.2.2 Spray Machine/Disk Maintenance

The high-speed machines that drive a disk atomizer are complex pieces of equipment and are subjected to extreme operating conditions. Several maintenance areas are critical to achieving acceptable operating factors. Vibration monitors are usually installed in order to detect an out of balance situation before serious damage occurs. Vibration can be caused by improper wheel balance, material build-up in the wheel, bottom guide bushing or cage wear that will ultimately allow shaft deflection. The plant must maintain the lubrication system of these machines including regular replacement of the oil, oil filter and cleaning of the oil line check valve. When slurry feeds are being used accumulation and ultimately pluggage of the feed tube, liquid distributor and the spray wheel can occur. Operating experience for the various materials being processed can be used to establish a regular cleaning schedule.

303.3 Heat Sensitivity

Spray drying is applicable to heat-sensitive products, such as milk powders and other foods and pharmaceuticals, because of the short contact time in the dryer hot zone. Because of the large total drying surface and small droplet sizes created, the drying rate is high, and the actual drying time in a spray dryer is measured in seconds. Furthermore, the liquid film on the droplet initially protects the solids from high gas temperatures. Inlet-gas temperatures may range from 398 to 1073°K (125 to 800°C), while the droplet temperature is essentially at the drying-air wet-bulb temperature, which is much lower. Surface moisture is removed rapidly, (typically in less than 5 seconds) but falling-rate drying takes much longer. The mean residence time of the gas in the chamber may range from 5 to 30 s. However, because of settling and mixing, the average residence times of particles can be substantially longer (as much as 60 s). The residence time distribution of particles is dependent on the mixing behavior of the gas and on the size distribution. The coarsest particles, which take the longest time to dry, fall more rapidly, and hence exit more rapidly from dryers with vertical flow. When conditions are adjusted to dry the coarsest particles, the finest may be overheated, and should be examined for degradation. The shape of the chamber affects residence time distribution (See Section 201.2).

The temperature to which the particles are exposed depends primarily on the inlet gas temperature and the extent of evaporative cooling. This in turn depends on the feed rate and water content. The maximum inlet air temperature is typically 10 to 20°F less then the melting point or char point of the solid in the feed solution. Even if the liquid spray does not contact the walls of the drying chamber, it is possible for dust to contact the wall even at a location above the spray nozzles. If the wall temperature is greater than the melting point of the solid, any solids contacting the walls will stick and melt forming solid buildup. (Table 303.6 footnote) Heat degradation also depends on particle size distribution, which depends on atomizing conditions. To find optimum or even satisfactory conditions for these variables requires exploring their effect. Systematic ways of doing this are described in Sections 303.4.1 and 502.1. To determine the product sensitivity to these operating conditions, take samples of the product from each experimental condition and later evaluate their product quality (such as color or flavor, dryness, flowability, density, particle size, and shape).

Other constraints on operating conditions other than heat degradation are thermal efficiency, productivity, and dryness of product. If adequate drying cannot be achieved at a low enough temperature, the liquid feed rate and the atomization conditions should be varied. Finer droplets will dry quicker, but will produce finer product particles. Lower feed rate will cool the hot gas less and allow use of a lower hot-gas inlet temperature. Conversely, if loss of product quality is not detected, atomization may be altered for a higher production rate. The effects of atomization are discussed under Particle Size in Section 303.4.2.

303.4 Physical Properties

The physical properties of spray-dried materials may vary considerably, depending on the inlet gas direction of flow and temperature, the degree and uniformity of atomization, the feed solids content and temperature, and the degree of aeration of the feed. (*Reference 808.1, 808.2*) The properties of the product usually of greatest interest are (1) moisture content, (2) particle size, (3) particle shape, (4) bulk density, and (5) dustiness.

303.4.1 Product Moisture

Product moisture is inversely related to the dryer gas outlet temperature. Dryer outlet temperature is the normal control variable for dryer operation. The product moisture is also affected by particle size distribution, atomizer operation, and gas flow distribution in the dryer. See Section 303.1.

303.4.2 Product Particle Size

The particle size is a function of atomizer operating conditions and also of the feed solids content, liquid viscosity, liquid density, and feed rate. In general, particle size increases with solids content, viscosity, density, and feed rate. Viscosity increases with solids concentration and with the inclusion of additives, and decreases with temperature. The feed rate increases with the scale of the equipment and with atomizer conditions.

303.4.2.1 Effect of Atomizer vs. Drying Conditions

Droplet size can affect residual moisture content and other properties such as product particle size. Therefore, tests should be done to determine the effect of atomizing conditions at suitable drying conditions. If nozzles are used for atomization, plan a series of tests with two or more nozzles of different orifice size. For each nozzle, check to see that drying is satisfactory; if not, vary the inlet gas temperature as described in Section 502.1. Then choose one preferred nozzle and vary the atomizer pressure. With a disc atomizer use two or more speeds. With the preferred atomizer settings determined, vary the inlet temperature to optimize capacity without degrading the product.

303.4.2.2 Feed Properties

Particle size is also affected by feed liquid composition. Increasing the liquid viscosity, e.g., by including additives, increases droplet size. Increasing liquid density by increasing concentration of dissolved or suspended solids also increases particle size. If feed composition can be varied, include these variables in your test plan. For example, try the effect of two or more feed concentrations on particle size at fixed atomizer conditions. If the results show a desirable effect, then conduct additional tests varying other conditions at the best feed concentration. At this point a factorial experiment plan may be required because of the number of variables involved.

303.4.3 Particle Shape

One of the principal advantages of spray drying is the production of a spherical particle, which is usually not obtainable by any other drying method. Spherical particles may be obtained from true solutions or from slurries, and may be produced by any of the previously described atomizers. This spherical particle may be solid or hollow, depending on the material, the feed condition, and the drying conditions. In general, aqueous solutions of materials such as soap, gelatin, and water-soluble polymers (which form tough tenuous outer skins on drying) will form hollow spherical particles when spray-dried. This is attributed to the formation of a case hardened outer surface on the particle, which prevents liquid from reaching the surface from the particle interior. Because of high heat-transfer rates to the drops, the liquid at the center of the particle vaporizes, causing the outer shell to expand and form a hollow sphere. Sometimes the rate of vapor generation within the particle is sufficient to blow a hole through the wall of the spherical shell.

If solid spherical particles are desired, the test should include low gas temperatures to reduce the rate of evaporation from within the droplet. The effects of feed composition (especially concentration and viscosity) can also be explored. The shape of particles, and the presence of hollow or broken spheres can be evaluated by microscopic examination.

303.4.4 Bulk Density

Bulk density is not as critical as moisture content and particle size for most products. (*Reference 808.1 & 808.2*) If bulk density is important, the bulk density of material from a spray dryer can be increased by the following operating changes:

- reducing droplet size,
- reducing inlet-air temperature,
- increasing air throughput,
- increasing air turbulence,
- employing countercurrent rather than co-current gas flow,
- effecting a wide range of size distribution from the atomizer.

Hollow spheres have a much lower density than solid ones, so conditions to produce solid spheres should be chosen. These variables should be included in the test plan if bulk density is important. Some effects of operating conditions on bulk density for a few materials have been reported (*Reference 808.8*) These variables also affect the product particle size and the achievable production rate or product dryness, which may be more important concerns.

303.4.5 Size Distribution/Dustiness

The size distribution of the product affects its dustiness, as well as other properties, such as flowability and bulk density.

303.4.5.1 Size Distribution

In general, the product size distribution will depend on atomizer design, liquid properties, and degree of atomization. See Section 303.2. Increased atomization energy (in the form of higher wheel tip speed, higher nozzle pressure, or higher nozzle air to feed ratio) will generally produce smaller average particle sizes. If the finest atomization possible is attempted, the particle-size range approaches a limiting condition and, regardless of the method of atomization, will be narrow. This is particularly true of pressure nozzles, in which uniformity of size increases with pressure. On the other hand, for the production of a coarse product with a high percentage of large particles, the method of atomization will have a large effect on the particle-size distribution. By careful design, centrifugal disks frequently can obtain production of uniform coarse particles.

303.4.5.2 Dustiness

A dusty product contains a high proportion of fines. This is caused by fine atomization or particle breakage after drying. Thin-wall hollow particles are susceptible to breakage during collection. Fine atomization and a high gas temperature contribute to high production rates in small drying chambers; but they also generate fine particles and thin-wall spheres. Attempting to increase production rate by these means can therefore lead to a dusty product. Therefore if dustiness is important, it should be one of the criteria for the test plan.

Some spray dryer designs include counter-current airflow, so that fines are carried out the top and coarse product settles to the bottom. Dustiness can also be remedied by agglomerating the particles after drying -- known as "instantizing."

303.5 Prediction of Capacity and Rate Effects

The examples in Sections 802, 803, and 804 should be studied to obtain guidance in planning the test. Observe in example 804, the effect of some variables can be estimated to obtain quantitative guidance for experimental studies, while other variables depend so much on the properties of the feed material that only trends can be predicted. For example, reducing the inlet feed moisture content will directly increase the powder production capacity, because most of the drying curve of most materials is constant at the wet end. A reduction of feed moisture from 1.5 to 1 kg/kg (60% to 50% moisture) will increase the production rate by nearly a factor of 1.5/1 (1.49 by Eq. 3). Since the same amount of water is to be evaporated, the other calculations in the example are not affected. Likewise it may be possible to increase the feed rate by increasing the inlet temperature, sufficient to provide enough heat for the increased evaporation, if the product is not too heat sensitive.

On the other hand, if the required moisture content of the final product in example 804 is reduced by half from 0.0526 to 0.0263, this will reduce the evaporation rate (Ev), since the drying rate will be lower at the dry end of the drying curve. It may be necessary to decrease airflow rate to lengthen residence time even more than proportional to moisture content reduction. This may require that the production rate be decreased more than the amount necessary to evaporate the small incremental amount of moisture, or the outlet temperature must be increased. These results can only be determined by experiment. The *direction* of changes can be predicted by discussions in Section 303, but quantitative changes cannot be predicted.

303.6 Summary

Table 303.6 summarizes the effect of some variables on product properties.

Table 303.6. Effects of Variables on Operation of Spray Dryers
Reference 808.1, p 277

Variable Increase	Factors Increased	Factors Decreased
Chamber Inlet Temperature	*Feed rate* and thus: product rate, particle size (*b*), product moisture content, chamber wall build-up (*d*)	Bulk density (*b*)
Chamber outlet temperature	Product thermal degradation	*Feed rate* and thus: product rate, particle size (*b*), product moisture content, chamber wall build-up (*c*)
Gas volume rate	*Feed rate* and thus: product rate, particle size (*b*), product moisture content, chamber wall build-up (*a*)	Residence time
Feed concentration	Product rate, Bulk density (*b*) Particle size (*b*)	
Atomizer speed Atomizer disc diameter *For stable lattices*	Bulk Density	*Particle size* and thus: Product moisture content Chamber wall build-up
For unstable lattices	Coagulation (*a*) and thus: Particle size Product moisture content Chamber wall build-up	
Atomizer vane depth Atomizer vane number	Bulk density (*b*)	Particle size (*b*) and thus: Product moisture content Chamber wall build-up
Atomizer vane radial length		*For unstable lattices* Particle size Chamber wall build-up
Feed surface tension	Bulk density (*b*)	Particle size (*b*)
Chamber inlet gas humidity	Product moisture content, Chamber wall build-up	

[a] This factor will only occur if a critical value of the variable is exceeded.
[b] Not for suspensions.
[c] Buildup occurs if the material is not dry. However, increasing outlet temperature can cause buildup if the material degrades.
[d] Inlet temperature should remain at least 20°C below melting point of the dry product to prevent sticking to hot dryer wall.

304 Data Requirements - Product Properties

The data that are required will depend on the specific objectives of the test. Some basic measurements required of all tests are listed below. Others must be added based on the test plan and its objectives. Product properties are reviewed in Section 303. There may be other properties such as color, taste and odor that are too diverse to cover in this procedure. Data units cited are S.I. and metric (kgs), but any consistent system may be used.

304.1 Data Requirements

All tests require sufficient measurements to determine a basic heat and material balance. This will include inlet air flow, temperature and humidity; feed rate and water or solids concentration of the feed; outlet air temperature and humidity; production rate (See Section 602.1), and product moisture content. The list follows:

Supply fan flow, m^3/s converted to kg/s
Spray dryer inlet gas temperature, °K (°C)
Spray dryer inlet gas absolute humidity, kg/kg
Feed rate, dry basis, kg/s
Feed moisture, dry basis, kg/kg
Spray dryer outlet gas temperature, °K (°C)
Outlet gas absolute humidity, kg/kg
Chamber product collection rate, kg/s
Product dust collection rate, kg/s
Chamber product moisture content, dry basis, kg/kg
Product dust collection moisture content, dry basis, kg/kg
Heat of reaction or crystallization, kJ/kg (kcal/kg)

304.2 Atomizer Data

Parameters to be measured include the dimensions and design of atomizer and drying chamber and the atomizer operating conditions. See examples 802 & 803 and reference 3, p 12-85.

Feed pressure, Newtons (kg/m^2)
Feed viscosity Pa-s (kg/m-s)
Feed density, kg/m^3 (g/cm^3)
Disk speed, s^{-1}
Atomizer dimensions, m (mm)
Chamber diameter, m
Chamber height, m

304.3 Complete Heat and Material Balances

If the accuracy of the tests is in question, then a more complete heat and material balance may be required. This includes ambient air humidity, heat loss from the dryer, product loss from the collection system, fuel use and heat value (for a balance around the heat source), and sensible heat in the feed and product:

Ambient air absolute humidity, kg/kg
Ambient air temperature, °K (°C)
Chamber, duct, and product collection equipment surface temperature, °K (°C)
Fuel consumption, kg/s
Fuel heating value, kJ/kg (kcal/kg)
(Alternative steam or electricity use), kg/s or kw
(Alternative steam latent heat value), kJ/kg (kcal/kg)
Feed temperature, °K (°C)
Spray dryer product temperature, °K (°C)
Dust collection equipment product temperature, °K (°C)

304.4 Peripheral Equipment Limitations

If dryer capacity is limiting performance, then measurements on peripheral equipment may be required, such as the exhaust fan, the product collection system, or the feed pumps. See example 804:

Exhaust Fan speed, s^{-1}
Exhaust fan flow, m^3/s
Exhaust fan static pressure, Pa (kg/m^2)
Exhaust fan power consumption, kw
Supply fan speed, s^{-1}
Supply fan static pressure, Pa (kg/m^2)
Supply fan power consumption, kw
Product collection equipment pressure drop, Pa (kg/m^2)
Product collection equipment outlet temperature, °K (°C)
Feed pump supply pressure, Pa (kg/m^2)
Feed pump outlet pressure, Pa (kg/m^2)
Feed pump power consumption, kw
Feed pump speed, s^{-1}

304.5 Product Quality Measurements

Some product quality criteria will usually be important, but this will depend on the application and should be established by the user. There may be other properties such as color, taste and odor that are too diverse to cover in this procedure. Methods of measurement will be those appropriate to the particular use or industry of the product. Examples of various quality criteria were discussed in Section 303:

Product bulk density, kg/m^3 (g/cm^3)
Product flowability
Product dustiness

304.6 Particle Properties

Microscopic examination will often be useful to understand how spray drying conditions are determining the properties of the product, especially the size distribution, presence of hollow particles, and fractured particles. See Section 303.4.

Product particle size distribution, μm
Product particle shape

305 Test Preparation

The following is suggested to prepare for the test.

305.1 Objective of tests

Describe the objectives in written words. List the important outcomes, such as capacity, efficiency, and product quality. Describe any product quality concerns.

305.2 Operating Variables

List the operating variables expected to affect the objectives. Identify design limitations of the spray dryer. Identify variables that can be controlled to effect the desired objectives, and list a range of operating conditions that can be controlled.

305.3 Test Data Sheet

In Table 805 (appendix), mark the items to be measured and those to be calculated.

305.4 Measurement Methods

List the intended methods to measure the important variables in the table.

305.5 Test Plan

Identify the starting conditions for the first test in the table. Also, define a program to vary these conditions to explore the objectives can be achieved. This step may need to be repeated in light of the results of initial experiments.

305.6 Planning Check List

During the test-planning period, the following preparations should be completed.

1. Obtain all burner, fan, heater, steam coil, and product collection equipment performance data from the manufacturers.
2. Obtain nozzle or atomizer performance curves, or make flow measurements at a range of conditions using actual feed material.
3. Determine fuel analyses and fuel-heating values.
4. Install all test instruments on the dryer or have them available at the test site for temporary use.
5. Provide all instrument connections and taps on the dryer installation, and assure safe access to all measurement points.
6. Test and calibrate all test instruments to assure they are in working order.
7. Prepare test data sheets; train personnel in the safe and proper use of all instruments and in the test procedures; and test and prove all laboratory analytical procedures for feed and product evaluations, *e.g.*, moisture determinations.
8. Inspect the equipment for mechanical soundness.

400 Methods of Measurement and Sampling

401 Gas Temperature and Humidity

401.1 Selection of Temperature Sensors

Exposed-junction or sheathed thermocouples connected to continuous indicator-recorders are preferred for rapid response and accuracy. Thermowell–enclosed thermocouples, and gas-filled temperature sensors also are acceptable provided time-constants are known. Dial thermometers are a poor third choice, and then only if scale division is sufficiently narrow to provide sensitivity. The use of glass thermometers for field tests on commercial dryers is an unsafe practice.

401.2 Installation of Temperature Sensors

Gas temperature instruments should be installed in gas streams having velocities in excess of 10 m/s (2,000 ft/min), to minimize radiation effects. If a temperature sensor is installed in the line of sight of a burner flame or near a high temperature steam or hot oil coil gas heater, a radiation shield may be necessary between the high temperature source and the sensor. Whenever such a heat source is 50°K hotter than the gas stream, possible measurement errors due to radiation and convection should be evaluated and the necessary compensations made.

401.3 Duct Temperature and Velocity Profiles

A thermocouple temperature is a point measurement and may not represent the mean temperature of a gas stream because, in any duct, the gas velocity and gas temperature may not be uniform across the full duct section. Before the performance test, all duct sections where temperature and velocity sensors are located should be thoroughly explored for temperature and velocity variations across the duct. This work should be done under drying operating conditions. A proper duct section profile will reveal the best location for a single-point measurement or reveal whether repeated profiles of the duct section also will be necessary to obtain accurate data during the actual performance test. Detailed description of techniques for gas velocity measurements are given in the Reference 3, Page 10-9 through 10-10.

401.4 Accuracy of Dry Bulb Temperatures

The expected accuracy for gas dry-bulb temperature measurements is ± 0.5% of the absolute gas temperature reading, °K.

401.5 Atmospheric Humidity

Atmospheric humidity is determined by comparing dry-bulb and wet-bulb temperatures using a sling psychrometer or by using an accurate aluminum-oxide hygrometer. Convert wet and dry bulb temperatures to atmospheric humidity using a psychometric chart for the local atmospheric pressure. See Section 602.3.

401.6 Dryer Exit Humidity

Dryer exit humidity should be measured downstream from all dry-type dust collection equipment, to sample the cleanest gas possible. Avoid vapor condensation on measuring devices by prewarming the sensor. Wet-bulb temperatures up to about 443°K can be measured by a wet-bulb thermocouple in clean gas. If the gas contains dust or exceeds 443°K in wet bulb temperature, a gas sampling method that cleans and cools the gas for dew point measurement will be needed. In this situation, standard sampling equipment is available and a manufacturer of gas moisture monitoring instruments should be consulted.

401.7 Accuracy of Gas Wet Bulb Temperature

The expected accuracy for gas wet-bulb or dew point temperature measurements is ±0.50% of the absolute gas temperature reading, °K.

402 Gas Flow

402.1 Installed Flow Meters

Very few commercial-sized continuous direct-heat spray dryers have permanently installed flow measurement nozzles, orifices, or venturi meters; therefore, for a performance test, it usually is necessary to employ portable pitot tubes or hot-wire anemometers.

> **NOTE:** Hot-wire anemometers must not be used in the presence of flammable or explosive vapors or dusts.

402.2 Inlet gas

402.2.1 Direct Measurements

Inlet gas (air) flow at ambient temperature can be measured by:

- Making a velocity profile across the total face area of an inlet gas duct or inlet gas filter using a hot-wire anemometer.
- Making a velocity profile across a convenient and reasonably straight, uniform section of inlet duct using either a pitot tube or a hot-wire anemometer.

402.2.2 Confirming Gas Flow Measurements

These gas flows measurements can be confirmed either by a calibrated pressure drop measurement across the heater or fan or by a heat balance.

402.2.2.1 Measurement By Pressure Drop

For steam-coil systems the pressure drop across the coil can be measured and compared to the coil manufacturer's published data showing air velocity versus pressure drop.

402.2.2.2 Measurement By Heat Balance

A heat balance comparing the fuel burned in the combustion gas heater to the gas temperature rise through the heater can confirm gas flow measurements. When making gas flow calculations based on the temperature rise attributable to fuel combustion, the lower heating value of the fuel must be used. The higher heating value, which is often cited by the supplier as the fuel's heating value, includes the latent heat of condensation of the water produced by the combustion of hydrogen in the fuel. This latent heat is not usable or recoverable in the dryer.

The steam condensate rate from the steam coils can be compared to the gas temperature rise across the coils. When measuring steam condensate flow from steam coils, a weight allowance must be added to account for condensate flash loss which occurs when condensate is released from a pressurized condensate system into an atmospheric-pressure condensate collector.

Electric and steam heaters and fuel burners should yield an efficiency of about 95%; the 5% loss is due to heater radiation losses and incomplete combustion.

402.2.2.3 Fan Performance Check

When a gas inlet system also has a supply fan, an additional check of airflow can be obtained by measuring fan speed, static pressure, and power consumed and comparing these to the fan manufacturer's data showing fan volume versus static pressure versus power.

402.2.2.4 Flow Measurement Accuracy

Agreement within ±5% between any two independent methods of inlet gas flow measurement is good agreement; ±10% is acceptable. The latter is the maximum measurement deviation that should be accepted without explanation as experimental error on a commercial-sized dryer.

402.3 Exit Gas Measurement

402.3.1 Direct Flow Measurement

For direct measurement of the dryer's exit gas flow, a pitot tube profile can be used of a clean duct section. Also, a hot-wire anemometer can be used if the duct gas temperature is below the maximum allowable by the anemometer specifications and flammable or explosive dusts are not present. If it is necessary to measure flow in a dusty gas stream, the Type "S" (Stauscheibe) pitot tube may be preferred over a regular pitot tube. The Type "S" is less susceptible to plugging and gives a higher differential pressure reading for any given velocity. (See References 11 and 12)

402.3.2 Effect of Wet Scrubbers

When the system uses a wet scrubber for recovery of dust, a clean gas flow measurement can be made downstream from the scrubber. In that case, the humidity change that occurs in the scrubber must be accounted for in order to determine the scrubber's inlet gas flow. The large fraction of water vapor in the scrubber exit gas significantly reduces the average gas density compared to dry air.

402.3.3 Recovery Equipment Pressure Drop

The powder collection equipment (cyclone or scrubber) pressure drop can be measured and compare to manufacturer's data, to confirm exit gas flow measurements.

402.3.4 Exhaust Fan Performance

The speed, static pressure, and power consumed by an exhaust fan can be measured and compared to the fan manufacturer's data, to provide a second confirmation of measured gas flow. In the exit gas from a dryer, note that the presence of a large fraction of water vapor and possibly carbon dioxide significantly affect the gas density. Gas composition as well as temperature must be known to determine gas density for use with fan curves and tables.

402.3.5 Exhaust Flow Accuracy

Agreement between the dryer's exit gas flow measurement and that indicated by cyclone pressure drop or fan performance is good if within ±10%; a deviation of ± 15% is the maximum acceptable without explanation.

403 Material Temperature and Moisture Content

403.1 Product Temperature Measurement

Probably the best way to measure the temperature of particulate solids is to place a representative sample in a closed, insulated container with a thermocouple immersed in the material to equilibrate in temperature with the material. If the material has either a low bulk density or a low thermal conductivity, it may be necessary to occasionally shake the container and immersed thermocouple to heat the thermocouple to the material temperature.

Thermocouples immersed in flowing streams of particulate solids usually yield unreliable data; the indicated temperature is likely to be closer to that of the entrained gas than to the temperature of the solids.

403.2 Product Moisture Content

During the performance test, material samples should be taken at frequent intervals, stored in gas-tight and pre-dried containers, and evaluated by a laboratory moisture test. On-line, continuous moisture meters may be acceptable for routine dryer control, but usually are not acceptable for a dryer performance test. Variations in moisture measurements that can be considered acceptable depend on material properties and the accuracy of the moisture test employed

Moisture samples of cyclone or bag collector product should be collected and analyzed separately from dryer product; following the same sampling precautions and analytical procedures.

404 Dust Flow Measurements

Material flow to the dust recovery system must be measured, collecting the recovered dust for a measured time period and weighing the increment collected.

405 Radiation and Convection Heat Losses

405.1 Estimated Heat Loss

Dryer heat losses depend on many factors, including dryer size, operating temperatures, dryer location, ambient conditions, and especially dryer insulation. New dryers usually have good insulation, and heat losses may amount to only 0.1 to 0.5% of the heat load. (Heat load is the sensible heat taken up by the solid, liquid, and vapor plus the latent heat of vaporization, as shown below in Eq. 10.) But many industrial dryers in operation have leaks and poor insulation, because these do not get repaired due to pressures to keep the unit in production. Many of these dryers were installed 30 to 50 years ago, when energy conservation had a low priority. In these cases heat loss may amount to 5 to 15%.

405.2 Measuring Heat Loss

The best way to account for radiation and convection heat losses is by making surface temperature measurements on all uninsulated surfaces and, as well as possible, on all insulated surfaces. The heat losses from these surfaces to the environment can be calculated using the data on pipe and duct heat losses, which are included in most engineering handbooks and texts dealing with heat transfer. See Section 603.1

It is helpful sometimes to determine overall heat transfer coefficients, which can then be used to estimate the heat losses when the dryer is operating at other temperatures (or for different ambient air temperatures). This can be done when no feed is being added to the dryer and the air heater is operating, by measuring the airflow through the dryer and the temperature at various locations through out the dryer system. The heat loss and the heat transfer coefficient can then be calculated. This method is also helpful to determine heat losses for the various dryer components (drying chamber, duct work, bag filter etc.).

Another useful diagnostic tool is Infrared Thermographic imaging. Hot spots indicative of inadequate insulation or hot gas out leakage can be located; this device will also indicate the surface temperature. It will also show cold spots, which may indicate an in leakage of ambient air. These images can provide a quick and effective means of locating problems and planning the needed repairs to maximize efficiency.

405.3 Outdoor Equipment

It is best to conduct performance tests on dryers installed outside, regardless of insulation quality, on dry, windless days to minimize heat losses and errors in heat balance calculations that might result from large heat losses.

If surface temperatures can not be measured, assume that on dry, windless days, the surface temperature of an uninsulated gas duct, within which the gas velocity is 10-30 m/s, is the average of the gas and ambient temperatures. Apply the same assumption to uninsulated cyclone or bag collectors. However, these assumptions should not replace actual surface temperature measurements when it is possible to make these arrangements.

406 Miscellaneous Measurements

406.1 Static Pressure

A spray dryer should always have at least one permanently installed static pressure indicator. It should be mounted either on the suction side of the process fan or on the direct-combustion gas heater if this heat source is used. In the case of fuel-fired heaters, instrument type and location must always conform to combustion equipment safety standards. Discussion of these standards, however, is beyond the scope of this procedure. During the performance test, other static pressure measurements can be made using temporarily-mounted water manometers or gauges, if this practice conforms to plant and product safety standards.

406.2 Location of Pressure Sensors

Care must be taken to ensure that static pressure instruments are installed in places remote from areas of high gas turbulence and points of gas flow direction change, such as in duct elbows and transitions. Be careful to obtain static pressure data from points of reasonably uniform gas velocity. This makes it possible to measure gas velocity at the same points and, if desired, calculate velocity head and total pressure.

406.3 Electric Power Measurements

Provisions should be made for safely measuring electric power consumption by the fan motors and electric gas heaters before the performance test. Qualified electricians should install the proper indicating or recording watt-meters on the motor and heater controls. Only a skilled electrician should use portable instruments designed for electric system troubleshooting during the performance test. If variable-speed drives are used, manufacturer's drive input power versus efficiency data should be available.

500 Test Procedure

501 Plant-Scale Test

Performance tests often fail, or yield inconclusive results, because of a desire to run full capacity equipment demonstrations too soon after a new plant start-up. The following startup and shake down items should be complete before a test:

- Operators are fully trained and familiar with the equipment,
- Start-up troubleshooting on all equipment upstream or downstream from the dryer is complete,
- Representative quality feed material can be produced at a rate great enough to supply the dryer.

It is not unusual for a year to elapse between an initial plant start-up and the time operators, equipment, and material are all ready for a valid full-capacity dryer demonstration. Carrying out a program of exploratory tests can speed up this learning process. When planning the timing of a performance demonstration, consider that startup of equipment may be delayed. See Section 700 for further discussion.

502 Exploratory Experiments

502.1 Production Capacity

Production capacity with acceptable product quality is the most important criterion of good operation for a particular application. It depends on several variables depending on the application, but the most important of these is the allowable outlet temperature, especially if the product is at all temperature sensitive, and the particle size. (See also Section 302.1 and 303.3)

Since particle size affects drying rate, both particle size and temperature should be investigated to determine capacity, as suggested in Section 303.4.2.1. This should be done in a systematic way. It may be most effective to choose an inlet temperature and vary the particle size by varying the atomizing conditions, and follow this by tests at other inlet temperatures. For each inlet temperature, adjust the feed rate to give an outlet temperature that will dry the product to the desired moisture content without product degradation. After tests have determined this feed rate, then vary the particle size by changing the atomizing conditions as indicated in 303.4.2.1.

With nozzle-equipped dryers, it may be necessary to shut down to change nozzles. In this case it may be more convenient to choose the atomizing conditions, set the feed rate by a heat balance calculation, and vary the inlet temperature, which will vary the outlet temperature. Other feed rates can be used to control outlet temperature and achieve desired product dryness. Then the nozzles can be changed to try other atomizing conditions. This can be an evolutionary experiment procedure.

A third method is to plan a factorial experiment. Whatever method is chosen, if the product is heat sensitive, sample the product at each condition to examine evidence of degradation, such as scorching. The procedure can be repeated changing other variables, such as the feed properties.

When testing a new heat-sensitive product in a full-sized dryer, start at conservative conditions that will not degrade the product. This starting point yields a small particle size, low inlet temperature, and low feed rate to give high outlet temperature for a dry product (but not so high as to damage the product.). To determine the feed rate to achieve a desired outlet temperature, base a rough heat balance (Section 603.1., Equation 10) on the fuel/air mixture and the heat of evaporation to dry the feed. Past experience with similar materials will help choose starting conditions. The inlet temperature may be 473°K or 80 to 90% of the design temperature, or even as low as 398°K for very heat-sensitive materials. At the lowest temperature, watch for undried product sticking to the dryer walls, or product agglomeration.

In pilot tests where cost of spoiled product is not important, another option is to start at a moderately high temperature of 473°K to 573°K (300°C). This will avoid the need to shut down and clean the dryer walls, which may happen if the lowest temperatures are used. If the product is not heat sensitive start at a temperature that the dryer can withstand, such as 773°K (500°C).

In either case, vary the temperature in steps, such as 50°. Run the dryer for 20 to 30 minutes at each setting and collect samples. Examine the samples to determine completeness of drying and product quality, and also check for product sticking to dryer walls.

502.2 Product Quality

Once these main operating conditions have been determined, other variables can be explored that may affect product quality. Section 303 discusses product quality interests or concerns that may govern a specific application and the variables that may be adjusted to achieve the desired quality results.

503 Preliminary Trial

When it appears that all facilities are ready for a performance test, a preliminary trial should be made. During the trial operate the dryer for a period of time at a rate as close as possible to the test rate. This should identify any major equipment deficiencies that, until corrected, could obviate the performance test. For example, if the gas heater does not attain the needed gas temperature, or the capacity of the exhaust fan is inadequate at the design fan speed, there is no purpose in proceeding with a test until the corrections are made.

503.1 Water Run

It is useful to start a preliminary test using clean water rather than the feed solution. This will locate leaks in the atomization system, verify the operation of the air heater at full load conditions and allow checking the operation of the temperature, pressure sensors and control loops, without consuming a lot of feed and producing off spec product. Then when switching to feed, attention can be devoted to the solids handling portion of the dryer system.

503.2 Dryer Operability

The preliminary trial will also provide an opportunity to inspect the dryer, atomizer and product collection equipment after a reasonable period of stable operation. Visual observation of the interior of the drying chamber can provide important insight into design flaws, improper dryer assembly or maintenance requirements. Product accumulation in a broad band at the rotary atomizer level may be caused

- by low drying gas flowrate,
- improperly adjusted turning vanes in the gas distributor, or
- inadequate atomization as indicated by coarse product and frequently caused by overflowing the atomizer wheel.

An asymmetrical accumulation near the atomizer may indicate a partial blockage of the gas distributor or some missing or twisted vanes.

Product build-up throughout a major portion of the drying chamber may indicate poor mixing of the droplets with the hot gas. This build-up could also be indicative of low gas inlet temperature, low flow, excessive feed rate or low feed solids.

Accumulation of moisture in the dryer discharge cone could be a sign of low outlet temperature, insufficient residence time or condensation caused by insulation failure or rotary valve blowby.

Problems in pressure nozzle dryers are somewhat easier to interpret. An accumulation in a localized area can usually be attributed to a single nozzle directed at that area. Another potential problem is use of an asymmetrical arrangement of nozzles resulting in poor gas/droplet mixing; this could have resulted from pluggage of individual nozzle(s). Nozzle misassembly or wear is the typical problem. Larger areas of accumulation can be a result of inadequate gas/droplet mixing. This can be caused by use of nozzles without a sufficiently wide hollow cone pattern to promote mixing of droplets with the hot gas.

The gas and product discharge ductwork should be inspected for product accumulation. A measurable build-up could be a result of low airflow, large particles, wet product or condensation. Also inspect the product collection equipment to ensure that cyclones are operating at the proper pressure drop and baghouses are pulsing properly. Check to ensure that product discharges at a consistent rate and that the pressure drop is steady; Variable pressure drop will result in inconsistent gas flowrate through the dryer system and may adversely affect the product quality.

504 Definitive Test

Once the tests have achieved a suitable setting of dryer operating conditions, the next step is to perform a definitive test, checking the results by material and energy balances. Section 602 describes in some detail the measurements and calculations that are required to perform such a definitive test.

504.1 Test Startup

To reduce the quantity of feed required for a test, it is useful to start the dryer system using clean water. Set the water feed rate to the same rate as the water fraction that is expected when using the actual feed solution. For example, if 1,000 kg/hr of feed containing 80 wt % water is to be fed to the spray dryer, start the spray dryer at a water feed rate of 800 kg/hr. Doing this will allow the temperature profile throughout the dryer to reach steady state while minimizing the quantity of feed required. Too high or too low outlet air temperature excursions during startup will not cause a problem. During this startup, damper settings, airflow rates and temperatures can safely be changed without the worry of material building up on the wall, or producing a large quantity of off-spec product, etc.

504.2 Running a Definitive Test in a Plant-Scale Dryer

When all preparations are complete, gradually increase the dryer feed rate to that specified for the test. Dryer operation at this feed rate should continue until all gas and material temperatures approach steady-state conditions. This will take at least one hour for a large dryer. At the end of this first period, record the relevant temperature, pressure, and flow measurements specified in Section 304. Uninterrupted operation should continue for a second period of at least one hour. Then, record a second set of temperature, pressure, and flow data. If this second set of data is essentially identical to the first, the test is complete. If not, hold for a third period of operation under steady-state conditions and essentially equal in length to the second period and record another set of measurements. This procedure should continue until two successive series of data, separated in time by at least one hour, are essentially identical.

505 Humidity and Moisture Measurements

Humidity measurements and material moisture samples should be taken at the same intervals as other data. If the dryer is believed to be operating on a borderline condition close to a maximum moisture specification, take material samples at more frequent intervals.

600 Computation of Results

601 Nomenclature

The following are the major items that are needed for heat and material balances, arranged in conceptual groups. For a more complete list see Section 801.

601.1 Variables

The following represent flows of heat and mass.

L = mass dry air flow rate, kg/s
P = Product or feed rate, on a dry solids basis, kg/s
Ev = Evaporation rate, kg/s
Q = Heat flow rate, kJ/s (kcal/s)

601.2 Properties

The following represent properties of material streams:

H = air absolute humidity, kg H_2O/kg dry air
C_p = heat capacity, kJ/kg °K
λ_0 = Heat of Vaporization of water at 273°K, 2501 kJ/kg. (597.3 kcal/kg)
w = Ratio of moisture to dry solids in feed slurry or solid product, kg/kg.
Vh = air humid volume, m^3/kg.
h = enthalpy of wet air stream, kJ/kg. (kcal/kg)

601.3 Subscripts

The following subscripts on variables or properties indicate position in the dryer.

i = inlet gas conditions
o = outlet gas leaving the dryer
f = feed conditions
p = product leaving the dryer.

602 Material Balances

602.1 Dry Solids

Often the product is collected in two fractions, a course fraction collected from the cyclone, and a finer fraction collected from a bag house, as shown in Figure 602.1. These fractions are combined to give the total product:

$$P_o = P_{dryer} + P_{cyclone} \qquad \textbf{(1)}$$

Masters (*reference 808.1,* p *410*) warns against using a material balance to detect losses of product in the dryer, because measurement errors usually far exceed such losses. Therefore,

$$P = P_o = P_f \qquad \textbf{(2)}$$

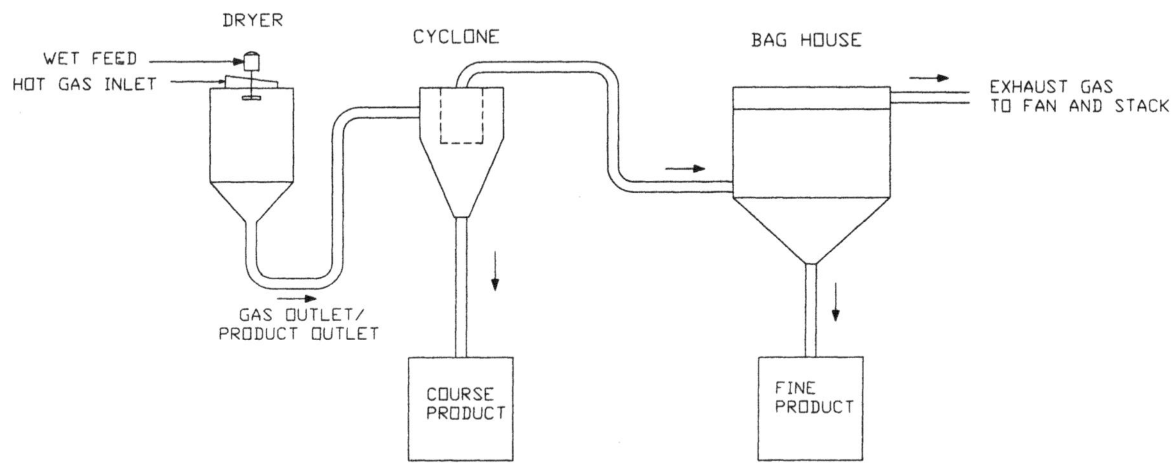

FIGURE 602.1. DRYER SYSTEM ARRANGEMENT

602.2 Moisture Balance

The moisture flows are defined as:

Moisture entering in the feed = $w_f P$, kg/h
Also, the wet feed rate = $(1+w_f)P$, kg/h
Moisture leaving in dried product = $w_o P$
Moisture entering in hot air = $H_i L$, kg/h
Moisture leaving in the dryer exhaust air = $H_o L$, kg/h

Note: If moisture content is expressed as percent of wet solids, $w_{\%}$, then

$$w = \frac{w_{\%}}{100 - w_{\%}} \text{, kg water/kg dry solids} \qquad (3)$$

Thus the dryer moisture balance is:

$$w_i P + H_i L = w_o P + H_o L. \qquad (4)$$

Or

$$(H_o - H_i) L = (w_i - w_o) P = \text{evaporation rate, } Ev\text{, kg/h.} \qquad (5)$$

This relation is useful to estimate any one condition from the other measured conditions or to check the measurement of both inlet and outlet conditions for example as described in the Section 602.6.

602.3 Psychrometric Chart Method

Use of a psychrometric chart for air-water systems is the most practical way to determine drying conditions from quantities that can be measured (especially wet- and dry-bulb thermometer readings). In addition, the chart can be used to make simplified heat balance calculations (*Reference 808.4 and Reference 808.1*), an alternative to use of Equation 10, below.

Psychrometric charts are available in *Perry's Chemical Engineers' Handbook, Spray Drying Handbook, Industrial Ventilation Manual,* and other standard references and textbooks. The charts in these sources usually are for sea level. For other elevations a dryer vendor may be a source or use "psychrometric chart" on an Internet search engine to get links to vendors of psychrometric chart software. The software typically will produce charts for any elevation and temperature range desired.

602.3.1 Reading Data from the Chart

The method of reading data from a chart is illustrated in Figure 602 *(Reference 808.3, p. 12-12.)* For example, using the psychrometric chart in Section 806 or Fig 12-3 in the *Reference 808.3 on p. 12-6*, to convert a dry-bulb of 80°F (26.7°C) and a wet-bulb of 67°F (19.4°C) to a moisture content:

- read up in the chart from 80°F on the bottom scale,
- read diagonally down the constant enthalpy line from the wet-bulb 67°F,
- from the intersection of these lines read across to the moisture scale on the right side, where moisture is 0.11 lb water/lb dry air (kg/kg).

Once such an intersection point is determined, you can also read off the relative humidity, dew point, specific volume, and enthalpy as shown in Fig 602.

Figure 602. Psychrometric Chart Use

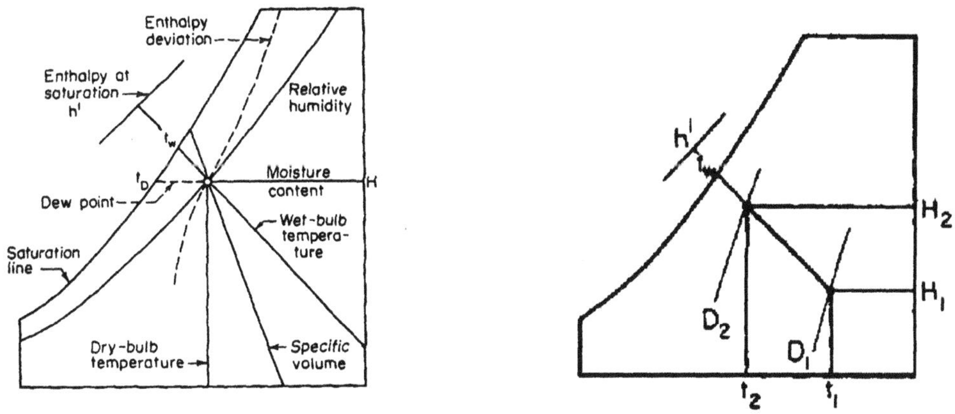

a. b.

602.3.2 *Simplified Heat Balance From The Chart*

Reference 808.3 also illustrates how to read results of various heating and cooling processes from the chart. For spray dryers, under conditions where heat loss and dilution air (the last two terms in the heat balance Equation 10) can be neglected, the heat balance is simplified and conveniently read off the chart. This is called the adiabatic case, since the heat to evaporate water from the product comes only from the sensible heat of the drying air, air enthalpy h_a of the air is constant. This includes the enthalpy of both the air and its moisture content, and is sometime called humid heat: (*Reference 808.4, p. 37)*

$$h_a = Cp_a\, T + H_a\, h_v \qquad \textbf{(6)}$$

where

H_a = moisture in air, kg/kg
h_v = enthalpy of water vapor, kcal/kg
T = temperature, °C

By following the diagonal adiabatic cooling line, along which h_a is constant, from the inlet conditions T_i to the outlet temperature T_o, a point on the chart is found representing the outlet conditions. The outlet air absolute humidity H_a can be read from the right scale of the chart (as illustrated in Figure 602 b.)

Example 802 shows how to apply this procedure to determine a simplified heat balance in a spray dryer.

602.4 Humid Air Volume

The specific humid air volume, needed to calculate flow velocities, is determined from the ideal gas law:

Specific Humid air volume = Vh m^3/kg of humid air

$$Vh = \left(22.4m^3 / kg\,mol\right)\left(\frac{1}{29kg / kgmol\,Air} + \frac{H_o}{18kg / kgmol\,WaterVapor}\right)\left(\frac{T^\circ K}{273^\circ K}\right)\left(\frac{1\,atm}{P\,atm}\right)$$

$$Vh = (0.00283 + 0.00456\,H_o)(T + 273), \qquad \textbf{(7)}$$

Where T may be T_o or T_i

Correct for pressure, if the plant elevation exceeds 300m. This quantity can be read from the psychrometric chart. The same method can be used for other combinations of gas and volatile liquid.

602.5 Total Air Volume

Evaluate the total air volume by multiplying Vh by the dry air mass flow:

$$V = Air\ volume = Vh\,L \qquad \textbf{(8)}$$

602.6 Air Material Balance

When airflow is measured by volume at the either inlet or outlet, L can be back-calculated from the above relation. If both flows are measured, compare the outlet with the value measured at the inlet. The difference should measure any substantial air leaks, although air leaks will normally be less than measurement errors, thus:

$$L_i = L_o + L_{leaks} \qquad \textbf{(9)}$$

In some cases equipment constraints may prevent measurement of L_i directly, in which case by neglecting leaks the volume flow measured at the outlet may be used to determine the air flow in the dryer.

603 Heat Balances

603.1 Enthalpy or Heat Balances on the Dryer

An enthalpy balance on the dryer gives *(Reference 808.1, p. 175)*:

$$Cp_{mixt}(T_i - T_o)L = (\lambda_0 + Cp_{vap}T_o - Cp_{liq}T_f)E_v + PCp_p(T_p - T_f) + Q_{los} + Cp_{aux}L_{aux}(T_o - T_{amb}) \qquad \textbf{(10)}$$

The left side of the equation gives the enthalpy given up by the hot air to dry the product. The first term on the right is the enthalpy to evaporate the water in the feed, plus the enthalpy to heat the water vapor to the outlet temperature, less the enthalpy in the liquid water in the feed (The initial conditions cancel out. The enthalpies are based on 0°C.) The evaporation rate is E_v. The next term is the enthalpy to heat the dry powder from the feed temperature to that of the product leaving the dryer. Q_{los} is the heat lost by convection from the surface of the dryer.

$$Q_{los} = UA\Delta T \qquad \textbf{(11)}$$

Where:

U = heat transfer coefficient,
A = area of the dryer surface,
ΔT = Temperature difference between the dryer surface and the ambient air.

For insulated chambers *U* is so small, the heat loss is usually negligible, and for uninsulated chambers *U* may amount to 20-90kJ/m^{2}h°K.

The last term is the enthalpy to heat any auxiliary dilution air supplied to cool a heat-sensitive product.

603.1.1 Heat of Reactions or Crystallization

In some cases where heat of reaction or crystallization are important, these must be added to right side of the above equation (10).

603.1.2 Back Calculation of Evaporation Rate from Enthalpy Balance

Equation (10) can be useful to determine the conditions for exploratory experiments. It can be solved for *E*v, and then can be used to determine the possible feed rate depending on the inlet and outlet temperatures and the given air rate, L. First, the moisture content of the feed relates P_i to Ev:

$(w_i - w_o)P_i = Ev$. Substituting from Equation 10,

$$Ev\left(1 + \frac{\rho Cp_P(T_P - T_f)}{(w_f - w_o)(\lambda_0 + Cp_{vap}T_o - Cp_{liq}T_f)}\right) = \frac{\rho Cp_{mixt}(T_i - T_o)L - Q_{los}Cp_{aux}L_{aux}(T_o - T_{amb})}{(\lambda_0 + Cp_{vap}T_o - Cp_{liq}T_f)} \qquad \textbf{(12)}$$

This can be solved for Ev, which in turn determines P_i.

603.1.3 Measurement Check

When both inlet and outlet conditions can be measured, equation (10) as a heat balance can be used to check the measurements. The results should agree within 10%. Comparison of terms in the equation also shows how the heat is supplied and used, and may facilitate adjustments for energy efficiency (Section 604). In particular, the effects of inlet and outlet temperatures are important, as is the solids concentration of the feed *(Reference 808.1, p. 111.)*

603.2 Heater Balances

603.2.1 Material Balances

603.2.1.1 Indirect Heaters

With indirect heaters, no heating medium dilutes the gas flow. Examples are steam, electric heaters, or gas-fired indirect heaters.

Usually ambient air is supplied to the heater with a blower, so that the air velocity and temperature can be measured in a duct downstream of the blower, or downstream from the heater, and L_i can then be calculated as described above in Section 602.5.

603.2.1.2 Direct-fired Heaters

603.2.1.2.1 Measurement Before Heater

If a good measuring point is not accessible after the heater or the air is too hot for a wet bulb, it may be possible to measure the air flow before the heater. If it is a direct-fired heater, then the primary combustion air flow and fuel rate must be measured and added to the total. In this case (Fig 603)

V_a = Primary air volumetric flow, m³/h
V_{comb} = Combustion air flow, m³/h

$$L_i = V_a/Vh_a + V_{comb}/Vh_{comb} + Fuel\ rate \qquad \textbf{(13)}$$

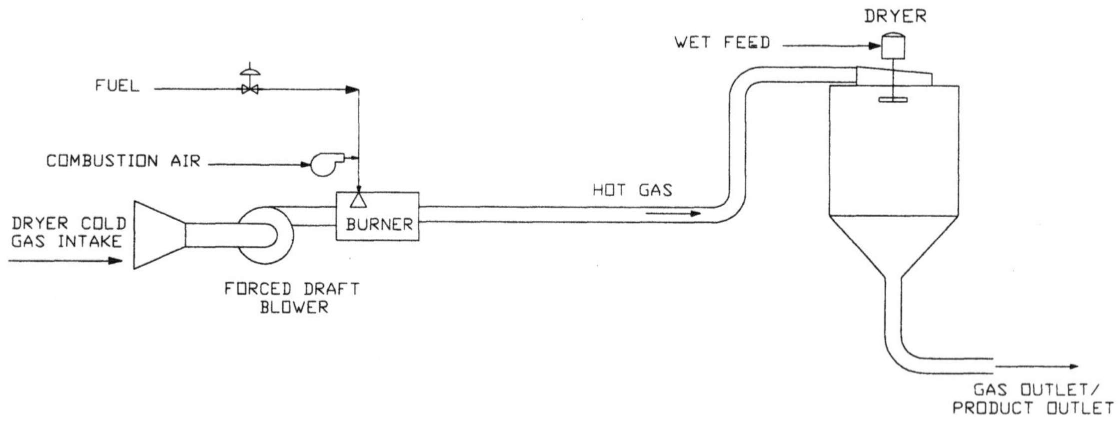

FIGURE 603. BURNER SYSTEM ARRANGEMENT

603.2.1.2.2 Determining Humidity and Airflow After Heater

If wet- and dry-bulb temperatures can be measured downstream of the heater, then calculate V_{hi} at this point. This allows L to be determined from velocity measurements. In *Perry's Chemical Engineer' Handbook* there are psychrometric charts that extend to dry-bulb temperatures of 250°F (120°C) and to wet-bulb temperatures of 130°F (55°C). Figure 806 is a psychrometric chart using a wide temperature range. In principle, higher temperatures could be measured, but it would be difficult to assure that the bulb remains thoroughly wetted, since the evaporation rate from the wick would be high. Therefore, some other means is needed to determine the dryer inlet moisture when using a direct-fired heater to reach higher temperatures.

One method is simply to ignore the water vapor from combustion products, because this may be unimportant unless the air is being heated to very high temperatures.

Another method is to measure the wet-and dry-bulb temperatures of the supply air going to the heater, and calculate the moisture added by the products of combustion , based on a heat balance. A detailed calculation requires the analysis of the fuel. *Reference 808.4* provides Table 603, which gives a simple method to determine moisture contributions for various fuels. The absolute humidity added by the fuel combustion is:

$$H_{comb} = (T_i - T_{amb}) * slope * 10^{-5} \qquad \textbf{(14)}$$

Then

$$H_i = H_{amb} + H_{comb} \quad \text{See Example 803} \qquad \textbf{(15)}$$

Table 603. Humidity Added by Combustion

Slope, (kg/kg, x 10^{-5})/°K

Reference 808.4

Heat Source	At 366°K	At 811°K
Indirect	0	0
No. 2 fuel oil	2.8	3.0
Propane	3.67	4.0
Natural Gas	4.14	5.04

603.2.2 Heat Balance and Fuel Consumption

603.2.2.1 Heat Required to Heat Inlet Air

The heat required to raise the inlet air from ambient temperature to the dryer inlet temperature is

$$Q_i = L_i Cp_{mixt}(T_i - T_{prim}) \qquad \textbf{(16)}$$

603.2.2.2 Indirect Heaters

603.2.2.2.1 Steam Heaters

For a steam coil heater, steam consumption is:

$$S = \left(\frac{Q_i}{\lambda_{stm}} \right) \text{kg/h} \qquad \textbf{(17)}$$

603.2.2.2.2 Electric Heaters

For an electric heater, the electric power required is:

$$Q_e = Q_i \, kw \qquad \textbf{(18)}$$

603.2.2.3 Direct-fired Heaters

Calculations typically assume 95% heater efficiency, due to assumed incomplete combustion. Measured fuel consumption should agree within 10% with calculated fuel consumption.

F = Fuel rate, kg/h
FLHV = Fuel lower heating value, kcal/kg. The lower value is used because it does not include the heat of condensing the water vapor, since condensation does not happen in a spray dryer.

$$F = Q_{fuel}/(0.95FLHV) \qquad \textbf{(19)}$$

The measured value F can be used to check the value of Q_i determined in Section 603.2.2.1.

604 Thermal Efficiency

604.1 Calculation

Thermal efficiency is defined (*Reference 808.1, p. 111*) as the ratio: Heat used in evaporation to Heat input.

$$Eff = \frac{(\lambda_0 + Cp_{vap}T_o - Cp_{liq}T_f)E_v * 100}{Q_i} \qquad \textbf{(20)}$$

It does not include the heat to raise the powder to the outlet temperature, since it is desired to minimize this heat load. Thermal efficiency can be calculated by substituting from equation (12) in Section 603.1.2, which gives the effect of the two main variables: temperature in the chamber, and moisture content of the feed.

$$Eff = \left(\frac{100}{Q_i}\right)\left[\frac{LCp_{mixt}(T_i - T_o) - Q_{los} - Cp_{aux}L_{aux}(T_o - T_{amb})}{(\lambda_0 + Cp_{vap}T_o - Cp_{liq}T_f) + \frac{Cp_p(T_p - T_f)}{w_f - w_o}}\right] \quad (21)$$

604.2 Optimization

Masters gives examples of optimization *(Reference 808.1, Table 3.6, p. 118)*. Optimization of thermal efficiency with respect to several variables can also be based on the heat balance Equation (21) in Section 604.1, as follows

604.2.1 Feed Solids

Masters *(Reference 808.1, p. 111)* notes that increasing the feed solids from 10 to 25% will reduce the heat load by 2/3, and increasing it from 25 to 50% will give another 2/3 reduction, assuming no decrease in the unit drying rate. The feed solids in some cases may be increased by more effective dewatering of the feed, or in other cases by mixing less water into the feed. The limitation on solids content is the rheology of the feed, which determines its spraying behavior. Preliminary spraying tests may be conducted to establish limits on feed solids content. However, the behavior of feed in the spray dryer may be complex, as discussed below in Section 701.

604.2.2 Dryer Temperatures

Thermal efficiency is also increased by increasing T_i and decreasing T_o. However, for heat-sensitive materials there is a limit to the upper temperature of operation. Although the particles are cooled by evaporation and theoretically do not exceed the wet-bulb temperature, in practice the finer droplets may dry first and then be heated to higher temperatures, so experiment is needed to determine the maximum allowable T_i.

604.2.3 Dew Point Limitations

Lowering T_o also increases thermal efficiency, but here care must be taken not to lower T_o below the dew point to prevent condensation. Masters *(Reference 1, p.97)* presents information to calculate the dew point. However, drying of the product can never lower air temperature below the dew point, because evaporation would then cease. Cooling below the dew point may occur by heat loss to the atmosphere through the walls of the product recovery equipment and exhaust stack, or on startup with cold equipment. It could also conceivably happen in a counter-current dryer due to sensible heat given up by a cold feed or by dilution effect with cold air *(Reference 1, Example p. 106)*

700 Interpretation of Results

Interpretation of results depends on objectives of the tests. This may be to measure capacity accurately, or thermal efficiency, or to explore performance, or troubleshooting. Heat and material balances help to evaluate accuracy of results.

701 Exploratory or Troubleshooting Tests

If the spray dryer performance does not meet expectations, either with regard to capacity or product quality or downstream product handling, it may be necessary to conduct a further series of tests. The difficulty is that there are so many variables that there may not be a clear path to reach the objectives. It has been estimated that there are 22 variables involved in spray dryer operation. For some industrial drying processes (e.g. drying in a tower spray dryer 80 ft. high by 16 ft. diameter) it may take two years of trial-and-error improvement efforts to reach a satisfactory performance, i.e. within 20% of expected capacity while making a satisfactory product. A satisfactory product should be homogeneous, with moisture content and particle size in the desired range, and with acceptable powder flow properties. Thus, a single test should not be expected to achieve the ultimate best results.

It may be possible to plan a systematic series of tests based on the discussion of the effect of variables given in Section 300. However, some of the variables are not well defined. For example, in detergent manufacture the raw materials must be made up into a slurry by dispersing particles in water, but slurries prepared by two operators will perform differently in the same spray dryer. This may be because mixing affects the viscosity, which in turn affects the flow properties through the nozzles. Furthermore, there is an interaction between the properties of the slurry and the design of the nozzles, including nozzle size and fluid pressure. A change in composition, such as the percent of active lauryl sulfonate ingredient, will affect the slurry behavior, as well as the handling of the powder product in downstream equipment. The same may be true for example when comparing ordinary coffee with decaffeinated coffee.

702 Material, Moisture, and Energy Balances

Example calculations are presented in Section 800 and follow the methods described in Section 600.0 Computation of Results. Nomenclature is presented in Section 801.0. The moisture and material balances should be checked to see that they agree within the recommended tolerances. Measured fuel consumption should be checked against that predicted by heat balance from airflow and temperatures. If these checks are in agreement, there is reasonable certainty that the test data are sufficiently accurate to measure spray dryer performance.

703 Dryer Heat Losses and Thermal Efficiency

703.1 Heat Losses

For well-insulated dryers, heat loss should be small. Cook's *(Reference 808.9 and 808.5, p 97)* calculations for 7-m diameter dryers with 50-mm insulation show a heat loss of 0.31% for low temperature (150°C in, 65°C out) dryers and 0.43% for high-temperature (540°C in, 150°C out) dryers. With 75-mm insulation, the results are 0.23 and 0.30%. If the insulation is poorly maintained so that there is 5% bare metal, the low-temperature dryer result is 0.60%; for 10% bare metal it is 0.90%; and for no insulation it is 12 to 18% heat loss. This assumes the loss from bare metal is 20 times that from insulation. The loss for smaller dryers is higher because of the higher surface-to-volume ratio. Thermal scans with an optical pyrometer are useful in detecting hot spots with insufficient insulation or hot gas leaks.

In any case, heat loss should not be used as a "fudge factor" to explain away a discrepancy in the testing heat balance. When the discrepancy is significant, the accuracy of the heat balance should be checked. If it is necessary to confirm heat losses, surface temperatures should be measured and a heat loss calculation should be performed. See Section 405 and 603.1 for measurement and calculation methods.

703.2 Thermal Efficiency

Thermal efficiency is defined in Section 604.2 as the portion of the total energy supplied to the air heater that is used to provide evaporative heat to the product. Thermal efficiency is dependent on heat losses, air leaks, heats of reaction and crystallization and operating temperatures in the dryer. However, thermal efficiency depends on many variables, so no hard and fast goal can be set for efficiency. For example, the inlet air temperature and humidity will affect the efficiency of a dryer operating at low temperature, but will be relatively unimportant on a high-temperature dryer. Another example is the approach to saturation of the exit air, defined in Example 803. With materials that are sensitive to heat and moisture there is a limiting approach to saturation that will not harm the product, and this limit can only be determined by experiment. Within this limit, the approach to saturation can be varied to affect thermal efficiency and capacity.

Although there are no rules, some examples can be given. With steam heat of air an overall efficiency of 40% has been found, requiring 1.2 to 1.8 kg steam per kg moisture evaporated. On the other hand, a direct-fired dryer operating at high inlet and low outlet temperature may have an efficiency of 80-85%.

The optimum efficiency is set by economics together with process limits. One reason is that the variables that affect efficiency also affect capacity, and there is often a separate requirement to produce at a certain capacity. If the economics of efficiency are important, then it may be worthwhile to explore the performance of the dryer by testing through a range of operating conditions that fall within allowable limits.

In many cases where high exhaust gas temperatures are required, gas-to-gas heat exchangers are used to preheat the incoming air with the outlet gas and thus improve thermal efficiency.

800 Appendix

The Appendix contains sections for Nomenclature (801), Examples (802, 803, & 804), Test Data Sheets (805), Psychrometric Chart (806) Index (808), and References (808).

The three examples demonstrate different methods of calculations based on available data. These also point out the versatility of using wet bulb temperatures and psychrometric charts in the calculations.

Example 802 demonstrates using psychrometric charts to make a simplified heat balance on the air stream of a dryer. Example 803 represents a spray dryer test where the air flow could not be measured, and was evaluated by heat and material balances from the product evaporation rate, using psychrometric charts. Example 804 represents a spray dryer test where both air and product properties were measured, so that heat and material balances could be used to check reliability of the data. It also uses equations rather than the charts to calculate a more complete heat balance.

801 Nomenclature

(The number in parentheses after = will convert the value in customary units into SI units.)

ΔT = temperature difference across dryer, or from dryer wall to ambient, °K (°C)
λ_0 = Heat of vaporization of water at 0°C, 2485 kJ/kg (594 kcal/kg)
λ_{stm} = latent heat of steam for indirect heater, 2272 kJ/kg (543 kcal/kg)
A = area of dryer chamber for heat loss, m^2
Cp = heat capacity, KJ/kg °K (kcal/kg °C). For dry solids, Cp may range from 0.25 to 0.4 kcal/kg.
Eff = dryer thermal efficiency, %
Ev = Evaporation rate, kg/s
F = fuel consumption, direct fired heater, kg/s for oil, or gas (m^3/s) = (ft^3/h)(x7.86×10^{-6})
FLHV = fuel lower heat value, for oil kJ/kg (kcal/kg) for gas kJ/m^3 (kcal/m^3), (Btu/ft^3)
h = enthalpy of wet air stream (Eq. 6), kJ/kg (kcal/kg)
H = absolute humidity, kg water/kg air
L = mass air flow rate, dry basis, kg/s
P = product or feed solids rate, on a dry solids basis, kg/s
Q = heat flow rate, kJ/s (kcal/s)
S = steam consumption, indirect heater, kg/s
Sa = adiabatic saturation, the ratio of absolute humidity of a gas to the absolute humidity when gas at the same temperature is saturated
slope = factor from Eq. 14, to obtain humidity contribution from temperature rise from direct-fired heater, kJ/kg °K. (kcal/kg °C)
T = temperature, °K (°C)
Twb = wet-bulb temperature, °K (°C)
U = Overall heat transfer coefficient for heat loss from dryer chamber, (kJ/s °K m^2) = (kcal/s-°C m^2)(x4.1868) or Btu/h °F ft^2
V = air volumetric flow, m^3/s, (ft^3/s)
Vh = humid volume of air at a given T, m^3/kg
w = Ratio of moisture to dry solids in feed slurry or solid product, kg/kg
$w_{\%}$ = moisture content of solids, wet basis, percent

Subscripts

The following subscripts on variables or properties indicate position in the dryer.

a = air
amb = ambient
comb = combustion airflow to heater
e = electric heater
f = feed conditions
i = inlet gas conditions
liq = liquid
los = losses
o = outlet gas leaving the dryer
p = product leaving the dryer
prim = primary airflow to dryer
s = solid
stm = steam
vap = vapor

802 Example: Evaluating Drying Capacity of Air Stream

(Reference 1, p 106)

A spray dryer is supplied from atmosphere with air containing 0.007 kg water vapor/kg air. The air is preheated to 422 °K. To achieve the desired moisture content in the product, the humidity of the air leaving the dryer is specified to be at 15% relative humidity. Calculate the theoretical quantity of moisture that can be removed in the dryer, and the temperature drop through the dryer. See Table 802, where given conditions are in bold type.

Since the air is assumed to be heated by an indirect heater, which adds no water vapor to the air, its absolute humidity remains 0.007 kg/kg as it enters the dryer. The maximum amount of moisture that can be taken up by the air is the amount to saturate the drying air. This can never be achieved, and a much lower humidity is usually specified, in this case 15%. to assure rapid drying of the product.

Use the psychrometric chart, Fig. 806. Locate the starting point from the bold print in Table 802 at 300°F, 0.007 kg/kg. Following the procedure illustrated in Fig 602.a, read diagonally up the adiabatic line to the saturation curve, then vertically down to the wet-bulb temperature T_{wb} = 106°F (314°K). From the same point on the saturation curve, read horizontally to the right scale, where H = 0.051 kg/kg. The maximum theoretical moisture take up is therefore 0.051 - 0.007 = 0.044 kg/kg.

For a practical drying condition, the relative humidity at the outlet is 15%. Read diagonally down the same adiabatic line from the saturation curve to intersect the curved line of 15% relative humidity. Read down from this intersection to the bottom scale, where T_o is found to be 167°F (348°K.) Also read horizontally from the same intersection to the left scale, where T_{dew} = 93°F (309°K.) This shows that this air could be cooled (without adding any water by evaporation) to 309°K before it is saturated, whereas the dryer outlet will be 348°K and will still have drying capacity. The difference between 309 and 348°K represents the relative humidity at the outlet. Also read horizontally across to the right scale, where H = 0.037 kg/kg. Then the practical moisture removal is 0.037 - 0.007 = 0.030 kg/kg air. These results we show in Table 802.

Making a heat balance by following the adiabatic line neglects heating of the product to its outlet temperature, and heat losses, and any dilution air, the last three terms in the heat balance Equation 10. Corrections could be applied for these terms, but we would need to know L, P, and w.

This example does not specify a measurement of these three terms. If two of these terms are known, the other can be calculated from the moisture material balance Equation 4. The evaporation rate could then be determined and compared with the rate calculated from the heat balance as derived in this example above. Subsequent examples will evaluate these terms.

Table 802. Drying Conditions from Psychrometric Chart

Station	**T** DryBulb	**T** Wet Bulb	**Absolute Humidity**
			H, kg/kg
Dryer inlet	**422°K** 149°C 300°F	314°K 41°C 106°F	**0.007**
Dryer outlet	348°K 75°C 167°F	314°K 41°C 106°F	0.037
Outlet at maximum moisture removal	314°K 41°C		0.051

803 Example: Dryer Test with Psychrometric Chart Heat Balance

This example was taken from Cook's article *(Reference 808.4)*. It assumes that no airflows can be measured, so they are computed by balances. This would be the case if ductwork is inaccessible, or too short for valid velocity measurements. Nevertheless, airflow could still be checked by fan performance curves.

Objective: calculate the airflow, heat load and outlet saturation condition, using psychrometric charts.

803.1 Known Inputs

Measured temperatures, °F. See Table 803

	T_{amb}	Twb_{amb}	T_i	T_o	Twb_o
°F	60	50	260	140	100
°K	298	283	400	333	311

Product rate is 175 lb/hr at 5% moisture. Feed contains 65% moisture.
Heater is direct fired with natural gas. Gas Fuel rate F = 0.0059 m3/s (750 ft^3/h)

803.2 Evaluate Product and Evaporation Rate

Convert percents to weight ratios:

$$P = \left(\frac{100-5}{100}\right)175 = 0.0209\,kg/s\;(166\,lb/h)$$

$$w_p = \left(\frac{5}{100-5}\right) = 0.0526\,kg/kg$$

$$w_f = \left(\frac{65}{100-65}\right) = 1.857\,kg/kg$$

From Equation 4, Section 602.2

$Ev = (w_i - w_o)\, P$ = *evaporation rate*
= (1.857 -0.0526) x 0.0209 = 0.0378 kg/s (300 lb water/h)

803.3 Ambient Air Humidity, Enthalpy, and Humid Volume

From the wet- and dry-bulb ambient temperatures in Table 803, read H, h, and Vh from the psychrometric chart Fig 12-2 in the *Handbook* (*Reference 808.3*) as described for Example 802. Vh may be read from steam tables, or computed from the ideal gas law Equation 7. For English units, replace 22.4 m^3/kg-mol by 359 cu ft/lb-mol.(at 32°F). If the installation is more than 300 meters (1000 ft) above sea level, then correct for the normal barometric pressure in Equation 7. Data is shown in Table 803.

803.4 Humidity Added by Fuel Combustion

The humidity added by the fuel combustion products can be calculated from Table 603, Section 603.2.1.2.2.

$H_{fuel} = (T_i - T_{comb}) \times 2.3 \times 10^{-5}$ for natural gas, kg/kg

$(260\text{-}60) \times 2.3 \times 10^{-5} = 0.0046$ kg/kg

$H_i = H_{prim} + H_{fuel} = 0.0055+0.0046 = 0.0101$ kg water /kg dry air

803.5 Dryer Outlet Conditions

At the dryer outlet, using T_o and Twb_o, read from the psychrometric chart Fig. 806 or Fig.12-2 from the *Handbook* (*Reference 808.3*), the values of H, h, and Vh shown in Table 803.

803.6 Dry Air Flow Rate

Compute the airflow from the evaporation rate Ev and the humid volume Vh. From Equation 4.

$Ev = (w_i - w_o)\, P = (1.857 - 0.0526) \times 0.0209 = 0.0378$ kg/s (300 lb/hr) water

From the left side of Eq. 4, $(H_o - H_i)\, L = Ev = 0.0378$ kg/s (300 lb/h)

$$L = \left(\frac{300}{0.0330 - 0.0101} \right) = 1.65\, kg/s\ or\, (13100 lb/h)$$

This is constant for each stage in the dryer.

803.7 Total Air Flow Rates

From Equation 8 the volumetric air flow rate for each station is VhL, with results shown in Table 803. If flow rates can also be determined by measuring velocities, then these can be compared to check results.

803.8 Burner Heat Input

The heat produced by the burner is calculated from Equation 19

$Q_{fuel} = F \, x \, FLHV$

$Q_{fuel} = 0.00590 \text{ m}^3/\text{s} \; x \; 37300 \text{ kJ/m}^3 = 220 \text{ kJs}$

$Q_{fuel} = 750 \text{ ft}^3/\text{h} \; x \; 1000 \text{ Btu/ ft}^3 = 750{,}000 \text{ Btu/h}$

This may be compared with the enthalpy change of the air as obtained from values in Table 803

$$Q_i = (h_i - h_{amb}) \, L = (74.2\text{-}20.2) \text{ x } 13100 = 207 \text{ kJ/s } (707400 \text{ Btu/h})$$

The difference between these two may represent incomplete fuel combustion, and the three other neglected terms in the heat balance, See Equation 10.

803.9 Check Driving Force For Evaporation

To check that conditions at the outlet provide a reasonable driving force for drying, evaluate the relative humidity there. From the psychrometric chart at the outlet temperature in Table 803 the relative humidity is 25%, which is in an acceptable range.

Some authors (Cook) feel that a more useful measure for drying is to compare the absolute humidity with the adiabatic saturation Sa, a ratio of absolute humidities at the same enthalpy. From the psychrometric chart, at the Twb_o of 100°F, read across to the right scale, where H = 0.0425 kg/kg. Then

$$Sa = \left(\frac{0.0330}{0.0425} \right) = 77.6\%$$

This is also considered in an acceptable range.

Table 803. Operating Data, Measured and Calculated

Station	Location	T	Twb	Absolute Humidity	Enthalpy	Humid Volume	Airflow Dry Rate	Airflow Volume	Product Wet	Product Moisture
		°F °K	°F °K	H, kg/kgDA	h, Btu/lbDA KJ/kg	Vh, ft^3/lbDA m^3/kg	L, lb/h kg/s	ft^3/min m^3/s	P, lb/h kg/s	w%
Ambient	Heater Inlet	**60** **288**	**50** **283**	0.0055	20.2 47.1	13.2 0.82	13100 1.65	2882 1.36		
	Heat Source			0.0046						
i	Dryer Inlet	**260** **400**		0.0101	74.2 173	18.4 1.15	13100 1.65	4017 1.90	475 0.0378	**65**
o	Dryer Outlet	**140** **333**	**100** **311**	0.0330	70.7 165	15.9 0.99	13100 1.65	3472 1.64	**175** **0.022**	**5**

Bold numbers are measured; others are calculated.

804 Example: Typical Dryer Test

This example follows the path of a more typical dryer test, where the main experimental variables are experimentally determined, and the dependent variables are calculated and compared by material and heat balances.

804.1 Experimental Data

The dryer configuration is shown in Figure 601. The experimental data are shown in bold type in Table 804. Other data in the table are calculated as shown below. Additional measured data not shown in the table are as follows:

Fuel consumption F = 0.0349 kg/sec No. 2 fuel oil
FLHV = 45258 KJ/kg (10816 kcal/kg)
Cp_p = 1.465 KJ/kg °K (0.35 kcal/kg °C)
Cp_{liq} = 4.187 KJ/kg °K (1.0 kcal/kg °C)
Cp_{vap} = 1.884 KJ/kg °K (0.50 kcal/kg °C)
Dryer is 3 m diameter by 6 m high

804.2 Humidities (H)

The absolute air humidities for ambient and dryer outlet stations are based on measured wet- and dry-bulb temperature readings, and are read from the psychrometric charts. The results are shown in Table 804.

804.3 Air Flows – Outlet

Humid volumes will be computed using Equation 7 (psychrometric charts could also be used). At the outlet:

$$Vh = (0.00283 + 0.00456H_o)(T + 273) = (0.00283 + 0.00456 \times 0.0535)(130 + 273) = 1.239 m^3 / kg$$

Air mass flows are determined from the measured volume flow in table 804 and the humid volume using Equation 3

$$L_o = V_o/Vh_o = 8.326 \text{ m}^3/\text{sec} / 1.239 \text{ m}^3/\text{kg} = 6.72 \text{ kg/sec}$$

The air flow volume is measured after the cyclone because the air is cleaner there than after the dryer chamber.

804.4 Air Flows – Inlet

At the inlet, the measured primary and combustion airflows of ambient air to the heater are summed:

$$Vh = (0.00283 + 0.00456H_{amb})(T_{amb} + 273) = (0.00283 + 0.00456 * 0.0117)(20 + 273) = 0.845 m^3 / kg$$

$L_{prim} = V_{prim}/Vh_{amb}$ =5.248 (m^3/sec) / 0.845 m^3/kg = 6.21 kg/sec dry air

$L_{comb} = V_{comb}/Vh_{amb}$ =0.423 (m^3/sec) / 0.845 m^3/kg = 0.5007 kg/sec dry air

Inlet air mass flow is the sum of the heater flows:

$L_i = L_{prim} + L_{comb}$ = 6.71 kg/sec

This agrees with the outlet mass flow calculated above to good accuracy.

Table 804. Dryer Conditions for Example 804

Variable	Units	Station				
		Ambient	Combustion Air	Dryer Inlet	Dryer Outlet	Dust Collector Exhaust
T	°C	20	20	230	130	128
	°K	293	293	503	403	401
Twb	°C	17.5			43.2	
	°K	290.5			316.2	
Humidity	Kg/kgDA	0.0117	0.0117	0.0176	0.0535	
Humid Vol	Vh m^3/kg	0.8448	0.8448	1.464	1.239	1.239
Air Flow	L kg/s	6.21	0.5007	6.71	6.69	6.69
Air Flow	V m^3/s	5.248	0.423			8.326
Product	Wet kg/s			0.417	0.158	0.0087
	Dry kg/s			0.1667	0.1503	0.0084
T_{prod}	°C			20	110	120
	°K			293	383	393
Moisture	w, kg/kg			1.5	0.0526	0.031

804.5 Dry Solids Balance

$P_o = P_{chamb} + P_{dust}$ from measurements
= 0.1503 + 0.0084 = 0.1587

P_i = 0.1667 from direct measurement

The agreement is only fair, and suggests some loss, perhaps on the walls of the equipment, or experimental error.

804.6 Moisture Balance – Evaporation Rates

Comparison of moisture evaporation from product and uptake by air determines the evaporation rate and checks its accuracy.

804.6.1 From product by Equation 5:

$Ev = (w_i - w_o)\,P$ = (1.5 - 0.0526) x 0.1503 + (1.5-0.031) x 0.0084 = 0.230 kg/s water

804.6.2 From Airflows and Humidities

804.6.2.1 Contribution of combustion water vapor

The dryer inlet humidity requires addition of combustion product water vapor to the ambient absolute humidity. Cook's Table 603 (see 603.2.1.2.2) relates the added moisture to the temperature rise:

$H_{fuel} = (T_i - T_{amb})(9^oF/5\,^oC)x\,1.55x10^{-5}$ for natural gas
= (230-20)(9/5)x $1.55\text{x}10^{-5}$ = 0.00585 kg/kg

$H_i = H_{amb} + H_{fuel} =$ 0.0176 = humidity at dryer inlet, kg/kg

804.6.2.2 From Air Moisture Uptake

By the left side of Equation 5

$Ev = (H_o - H_i)\,L$ = (0.0535 - 0.0176) 6.71 = 0.241 kg/s

Agreement is good with 804.6.1.

804.7 Heat Balances: Heater Duty and Fuel Consumption

The fuel consumption was measured as 0.0349 kg/s fuel oil. The energy produced by the fuel is

Q_{fuel} = F x FLHV = 0.0349 x 45,272 kJ/kg = 1580 kJ/s (377 kcal/s)

An energy balance relates this to the heating of inlet air based on air temperature and flow measurements:

$$Q_i = L_i\,(Cp_{air} + Cp_{vap}\,H_i)(T_i - T_{amb})$$
$$= 6.71(1.004 + 2.09 \times 0.0176)(503\text{-}293) = 1466 \text{ kJ/s}$$
$$= 6.71(0.24 + 0.50 \times 0.0176)(230\text{-}20) = 350 \text{ kcal/s}$$

The ratio of Q_i / Q_{fuel} = 1466/1580 = 92.8% is primarily attributed to the burner efficiency, and is within a reasonable range. If this ratio were smaller, one might look for an error. The most likely cause is error in measurement of the inlet air flow, which depends on making an accurate scan of velocity over the duct cross-section. See *Perry's Chemical Engineers' Handbook, Reference 808.3 p. 10-9 to 10.* There could also be fuel measurement error.

804.8 Heat Balance Over Dryer

The heat balance is derived from Equation 10. Each term will be separately considered.

804.8.1 Heat solids to product temperature (Third term in Equation 10.)

$$Q_p = P_{chamb}Cp_p\,(T_{o\,chamb} - T_{amb}) + P_{dust}Cp_p(T_{o\,dust} - T_{amb})$$

$$= 0.158 \times 1.465\ (110 - 20) + 0.0087 \times 1.465\ (120 - 20)$$
$$= 23.9 \text{ kJ/s } (5.28 \text{ kcal/s})$$

804.8.2 Heat to evaporate and heat water (Second term in Equation 10)

$$(\lambda_0 + Cp_{vap}\,T_o - Cp_{liq}\,T_f)\,Ev = (594 + 0.50x130 - 1.0x20)\,Ev = 639\,Ev \text{ kcal/s.}$$
$$= (2487 + 2.09x130 - 4.19x20)\,Ev = 2675\,Ev \text{ kJ/s}$$

From 804.6.2 above, an average value of Ev is 0.235 kg/s, so this term = 628 kJ/s (150 kcal/s)

804.8.3 Heat loss from dryer surface (Equation 11)

Assume that U = 0.14 kJ/m^{2}s/°K (5 kcal/m^{2}h°C)

$$Q_{los} = U\,A\,(T_o - T_{amb}) =$$
$$5.0\frac{kcal}{m^3 h^\circ C} * \pi * \left(\frac{3}{2}\right) * 6 * (130 - 20) = 15500\,kcal/h = 18\,kJ/s\,(4.3\,kcal/s)$$

The sum of the three terms in this section 804.8 is 5.28 + 150 + 4.3 = 670 kJ/s (160 kcal/s).

804.8.4 Heat Delivered by Drying Air (Left term of Eq. 10.)

$Q_i = L_i\,(Cp_a + Cp_{vap}\,H_i)(T_i - T_o)$
$= 6.71(0.24 + 0.50 \times 0.0176)(230\text{-}130) = 167$ kcal/s
$= 6.71(1.004 + 2.09 \times 0.0176)(503\text{-}403) = 699$ kJ/s

The agreement with the sum of three terms [690 kJ/s (160 kcal/s)] in Sec. 804.8.3 is satisfactory.

804.8.5 Heat Lost in Exhaust Air

Considering this term will close the heat balance. The heat carried out of the dryer by the exhaust air is:

$$Q_{exhaust} = L\,(Cp_a + Cp_{vap}\,H_o)(T_o - T_{amb})$$

$$= 6.71(0.24 + 0.50 \times 0.0535)(130\text{-}20) = 197 \text{ kcal/s}$$
$$= 6.71(1.004 + 2.09 \times 0.0535)(503\text{-}293) = 825 \text{ kJ/s}$$

The sum of $Q_{exhaust}$ (from 804.8.5) and the sum of three terms to dry the product (from 804.7) should equal Q_i , the heat supplied by the inlet air. 197 added to 160 gives 357, in agreement with the heat supplied by the inlet hot air, 350.

804.9 Thermal Efficiency

Thermal efficiency is defined by Equation 20:

$$Eff = \frac{\left(\lambda_0 + Cp_{vap}T_o - Cp_{liq}T_f\right)E_v * 100}{Q_i}$$

$$= (594 + 0.50\text{x}130 - 1.0\text{x}20)0.235/350 = 43\%$$

804.10 Adiabatic Saturation

This is calculated by the method in Section 803.9. From the psychrometric chart, at the outlet wet bulb temperature of 43.2°C, H = 0.06 kg/kg; from Table 804, H_0 = 0.0535. Then S_a = 0.0535/0.060 = 89%. This is a relatively high degree of outlet air saturation, and it does not suggest trying to improve thermal efficiency by lowering the outlet temperature further, since this would decrease the production rate.

805 Test Data Sheet

Table 805 is a useful form for entering data. In the table, mark the items to be measured and those to be calculated.

Table 805. Dryer Conditions for Test Plans

Station	T	Twb	Humidity	Humid Vol	Air Flow	Air Flow	Product	T_p	Moisture
	°C °K	°C °K	H kg/kg	Vh m^3/kg	L kg/sec	Vol m^3/s	P, kg/s Wet Dry	°C °K	w, kg/kg
Ambient									
Combustion Air									
Dryer Inlet									
Dryer Outlet									
Dust Collector									

806 Psychrometric Chart

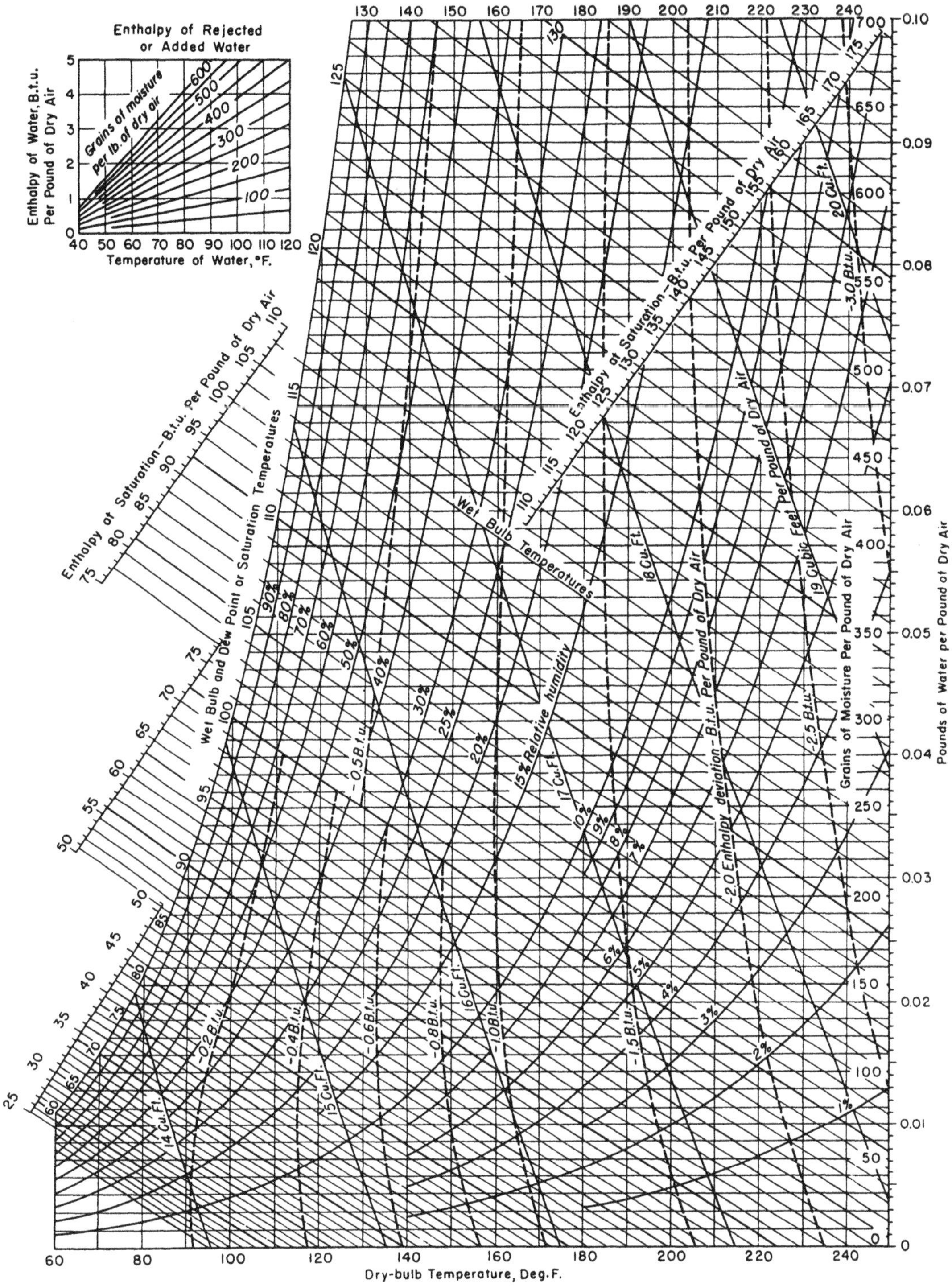

Reprinted with permission from *Perry's Chemical Engineers' Handbook*, 7th Edition, 1997, R.H. Perry and D.W. Green, Editors, The McGraw-Hill Companies.

807 Index

Acceptance Test 101.4, 302.4, See also Definitive Test
Accuracy 401.4, 401.7,402.3.5, 402.11, 603.1.3
Airflow 402, 804.3
Airflow Patterns 201.2
Anemometer, see Flow
Atomization 201.4, 201.3, 202.1, 303.2, 303.4, 304.2

Calculations see Computations
Capacity 301.1, 302.1, 303.5, 501, 502.1
Chamber Design 201.1
Cleaning and Inspection 301.9
Combustion, humidity added by, 603.2.1
Computations 301.6, 600
Consistency Test 302.2.6
Controls 301.4
Cyclone Flow Measurement 302.2.9

Data to Be Collected 302.2.7, 304, 305
Definitions 202
Definitive Test 504
Density 202.1, 303.4.4
Dew Point 202.1, 604.2.3
Disks, Spinning 303.2.2
Dryers
- Commercial-Sized 101.2, 501
- Configuration Chart 602.3, 802
- Design of 201, 301.1, 302.1
- Operability 503.2
- Pilot-Sized 101.2, 101.4.2
- Performance Data Tables 301, 302
- Variables Effect Table 303.6

Duct Measurements 401.3
Dustiness 303.4.4.5

Economic Optimum 302.2.2
Efficiency, Thermal 202.1, 302.2.2, 402.3, 604, 703.2, 804.9
Electric Power 406.3
Environmental 301.8
Equipment, Peripheral 302.1.2
Error, see Accuracy
Evaporation Rate 603.1.2, 803.2, 804.6
Example Calculation 802, 803, 804
Exploratory Experiments 101.4, 502.7.1
Explosions 301.7.2
Extrapolate Results 302.4.1

Factorial Test Plan 301.7.1, 502.1
Fans 201.7, 302.2.10, 402.3.4, 402.2.2.3
Feed Properties 303.4, 604.2
Flow Measurements 302.2.11, 401.3, 402
Fuel Measurement 302.2.8, 603.2

Heat Balance 202.1, 302.2, 304.3, 402.2.2.2, 602.3.2, 603, 603.1.2, 603.2.2, 802, 803, 804.7, 804.8.
Heat Losses 405, 703.1, 804.8.3.
Heat of Reaction 603.1.1
Heat-Sensitive Materials 201.2, 303.3
Heater Balance 603.2, 702
Heater Coil Flow 302.2.12, 402.3
Heating Method 201.5
Heating Value of Fuel 402.3, 603.2.2., 803.8, 804
Humidity 600
- Calculation from Chart 602.3, 802
- Contribution from Burners 603.2, Table 603, 804.6.2
- Effect on Product 802

Instrumentation 301.4, 305

Material Balance 302.2, 304.3, 602, 602.6, 803, 804
Measurements 302.2.4, 400, 403. 406, 804
Moisture Balance 602.2
Moisture Content 202.1, 303.1, 303.4.1, 403, 602.2

Nomenclature 601, 801
Nozzles 303.2.1

Objectives 301.1
Organization 301.2

Particle Size 303.4.2
Peripheral Equipment 302.1, 304.4
Pitot Tube, see Flow
Preliminary Test 503
Pressure, Static 406.1, 406.2
Pressure Drop 201.7.
Product Properties 201.3, 302.3, 303.4, 304, 304.5, 304.6, 502.2
Product Recovery 201.6, 302.2.5
Production Rate, see Capacity
Psychrometric Chart 602.3, 802, 803, 806

Rcasons for Tcst 101.4
Residence Time 202.1, 303.3

Safety 301.7.2
Sampling 400, 403.2
Scrubber 402.3.2
Sensitivity, Heat 303.3
Sensors, Temperature 401, 403.1
Shape of Particles 303.4.3
Size Distribution 303.4.5
Steady State 302.2.3
Steam Coil Flow 302.2.12, 402.2

Temperature Measurements 401.1, 401.3, 401.4, 403
Temperatures 401, 604.2.2
Test Preparation 305, 500
Test Planning 300
Test Plan 301.7
Test Operation 504
Thermocouples 401
Time, Residence 303.3
Timc, Drying 303.3
Troubleshooting 101.4

Variables 102, 301.7.1, 303.6, Table 303.6, 304, 601, 801
Variables, Multiple 301.7.1
Velocity Profile 402.2
Volume of Air 602.4, 602.5

808 References

808.1 Masters, Keith, *Spray Drying Handbook*, Longman Scientific and Technical, Essex, England, 5th ed., 1991.

808.2 Moyers, C.G., "Spray Dryers," *Perry's Chemical Engineers' Handbook*, McGraw-Hill Book Co., 7th ed., 1997, pp12-81 to 12-90.

808.3 *Perry's Chemical Engineers' Handbook*, R.H. Perry and D.W. Green, eds., The McGraw-Hill Companies, 7th ed., 1997, pp 3-87 to3-88, 10-9 to 10-10, 12-3 to 12-12, & 12-81 to 12-90.

808.4 Cook, Edward, "Drying: As Much Art as Science," *Chemical Processing*, Dec 1999, pp 36-39.

808.5 Cook, Edward M. & DuMont, Harman D.; *Process Drying Practice*, McGraw-Hill, Inc., 1991; ISBN: 0-07012462-0.

808.6 Walas, Stanley M., *Chemical Process Equipment: Selection and Design*, Butterworth Publishers, 1988, pp 268 to 277.

808.7 Kuehl, R.O., *Statistical Principles of Research Design & Analysis*, 2 ed, 2000, Duxbury Thompson Learning, Pacific Grove, CA.

808.8 Duffie and Marshall, *Chemical Engineering Progress*, **49**, pp 417-480, 1953.

808.9 Cook, Edward, E-mail to Dr. R. H. Snow and J.F. Hasbrouck, March 12, 2001.

808.10 Fikova, I. and Mujumdar, Arun S., "Industrial Spray Drying Systems" in *Handbook of Industrial Drying*, Vol 1, 2nd edition; Editor Arun S. Mujumdar, 1995, Marcel Dekker, NY, pp 263-307.

808.11 International Standard Organization, Standard #ISO 3966: 1977, Measurement of Fluid Flow in Closed Conduits – Velocity Area Method Using Pitot Static Tubes.

808.12 International Standard Organization, Standard # ISO 7194: 1983, Measurement of Fluid Flow in Closed Conduits – Velocity Area Methods of Flow Measurement in Swirling or Asymmetric Flow Conditions in Circular Ducts by Means of Current Meters or Pitot Static Tubes.

809 Tables and Figures

Page 7. Figure 201, Typical spray dryer configuration. Autocad drawing by Hasbrouck Engineering. Reprinted with permission.
Page 7. Figure 202, Exploded view of a spray dryer system. Swenson Technology, Inc., Reprinted with permission.
Page 13. Table 301, Performance data of spray dryers, Reprinted from *Chemical Process Equipment: Selection and Design*, 1988, Stanley M. Walas, Butterworth Publishers. Reprinted with permission from Elsevier.
Page 14. Table 302, Performance of a dryer 18 ft diameter x 18 ft high with a spray wheel and airflow of 11,000 cfm at the outlet, Reprinted from *Chemical Process Equipment: Selection and Design*, 1988, Stanley M. Walas, Butterworth Publishers. Reprinted with permission from Elsevier.
Page 20. Table 303.4, Particle diameters, densities, and energy requirements, Reprinted from *Chemical Process Equipment: Selection and Design*, 1988, Stanley M. Walas, Butterworth Publishers. Reprinted with permission from Elsevier.
Page 21. Figure 300, One fluid spray nozzle with swirl chamber. Spraying Systems Company. Reprinted with permission.
Page 27. Table 303.6, Effects of variables on operation of spray dryers, Reprinted from Keith Masters, *Spray Drying Handbook*, Longman Scientific and Technical, Essex, England, 5th edition, 1991.
Page 41, Figure 602.1 Dryers system arrangement. Autocad drawing by Hasbrouck Engineering. Reprinted with permission.
Page 43. Figure 602, Psychrometric chart use, *Perry's Chemical Engineers' Handbook*, 7th Edition, R.H. Perry and D.W. Green, Editors, The McGraw-Hill Companies, 1997. Reprinted with permission.
Page 46. Figure 603, Burner system arrangement. Autocad drawing by Hasbrouck Engineering. Reprinted with permission.
Page 47. Table 603, Humidity added by combustion, Cook, Edward, "Drying: As Much Art as Science," *Chemical Processing*, Dec 1999, pp 36-39. Reprinted with permission.
Page 55. Table 802, Drying conditions from psychrometric chart, Cook, Edward, "Drying: As Much Art as Science," *Chemical Processing*, Dec 1999, pp 36-39. Reprinted with permission.
Page 60. Table 804, Dryer conditions for Example 804, Cook, Edward, "Drying: As Much Art as Science," *Chemical Processing*, Dec 1999, pp 36-39. Reprinted with permission.
Page 64. Table 805, Dryer conditions for test plans, Cook, Edward, "Drying: As Much Art as Science," *Chemical Processing*, Dec 1999, pp 36-39. Reprinted with permission.
Page 65. Figure 806, Psychrometric Chart, *Perry's Chemical Engineers' Handbook*, 7th Edition, R.H. Perry and D.W. Green, Editors, The McGraw-Hill Companies, 1997. Reprinted with permission.

www.ingramcontent.com/pod-product-compliance
Ingram Content Group UK Ltd.
Pitfield, Milton Keynes, MK11 3LW, UK
UKHW051507290926
15471UKWH00008B/1480

9 780816 909254